Long Way Home

The true story of an Australian lad's survival during the Japanese invasion of Rabaul in World War Two

C. Killen

LONG WAY HOME

Clarence (Tal) A Killen

First edition © Tal Killen 1998

Second edition © Cheryl Killen 2021

All rights reserved

Title: Long Way Home

Author: Clarence (Tal) Killen (1925–1999)

ISBN: 978-0-6450744-0-6

Set in Century Schoolbook 11.5

This book was written by Tal Killen in 1993.
As a lasting memory of Tal's life, the Killen family produced this edition in 2021.

Cheryl would like to acknowledge the kindness of John and Gwen Huth, who generously offered their time and editing expertise to help publish this second edition. Tal met and made friends with John and Gwen when he sought an opinion from Gwen on his manuscript all those years ago.

CONTENTS

INTRODUCTION i
PREFACE.. ii
THE BEGINNING 1
ADVENTURE.. 4
THE MEETING 6
KNOWLEDGE GAINED 9
VOYAGE TO RABAUL.. 15
ENLISTMENT... 33
THE LONG MARCH BEGINS.. 47
TOL PLANTATION 68
HIGHLAND JUNGLE 78
MURDER AVENGED 91
HOSKINS PENINSULA 100
GOING HOME.. 109
AFTERWORD 113

INTRODUCTION

This is a factual account of the author's experiences endured during the war in New Britain in 1942. For fifty years, feelings of guilt, and emotion have overwhelmed the author, and he was unable to tell his story to his family.

Hopefully, now, the author has found it less difficult to converse with his family about his wartime experiences.

The author believes that the reader, and in particular, his family, will gain a valuable insight into the awesome encounters witnessed and decisions made during war, and have an understanding of the lifelong affect upon all those who participate in war.

It is written in the third person in order that it may be seen through the reader's eyes and is dedicated to the memory of a loyal caring friend.

PREFACE

It had been an oppressively hot night and he'd slept rather fitfully, his sleep marred by dreams.

In the darkness before dawn a slow movement near his face suddenly brought those dreams into half reality. His body galvanised into instinctive reaction jerked violently away and an oath of fearful rage was torn from his dry throat. Trembling uncontrollably as reality slowly revealed itself, he felt a soft reassuring hand caress his sweating brow, just like a thousand times before. As he'd done a thousand times before, he eased himself from the bed, donned his gown and moved outside in the cold, pre-dawn darkness, angry with himself, to sit with his memories. His wife of forty years shed her tears for him as she lay alone.

He watched the brilliant morning star pale as dawn heralded day, and prayed to himself that one day he may be relieved of the memories of half a century gone.

At sixty-seven years he studied his frail wrinkled hands, felt the weakness in his body and reflected on the strength and vitality that was once his and had carried him through hell and back.

His thoughts took him back to half a century ago.

"GREATER LOVE HATH NO MAN

THAT HE LAY DOWN HIS LIFE FOR A FRIEND."

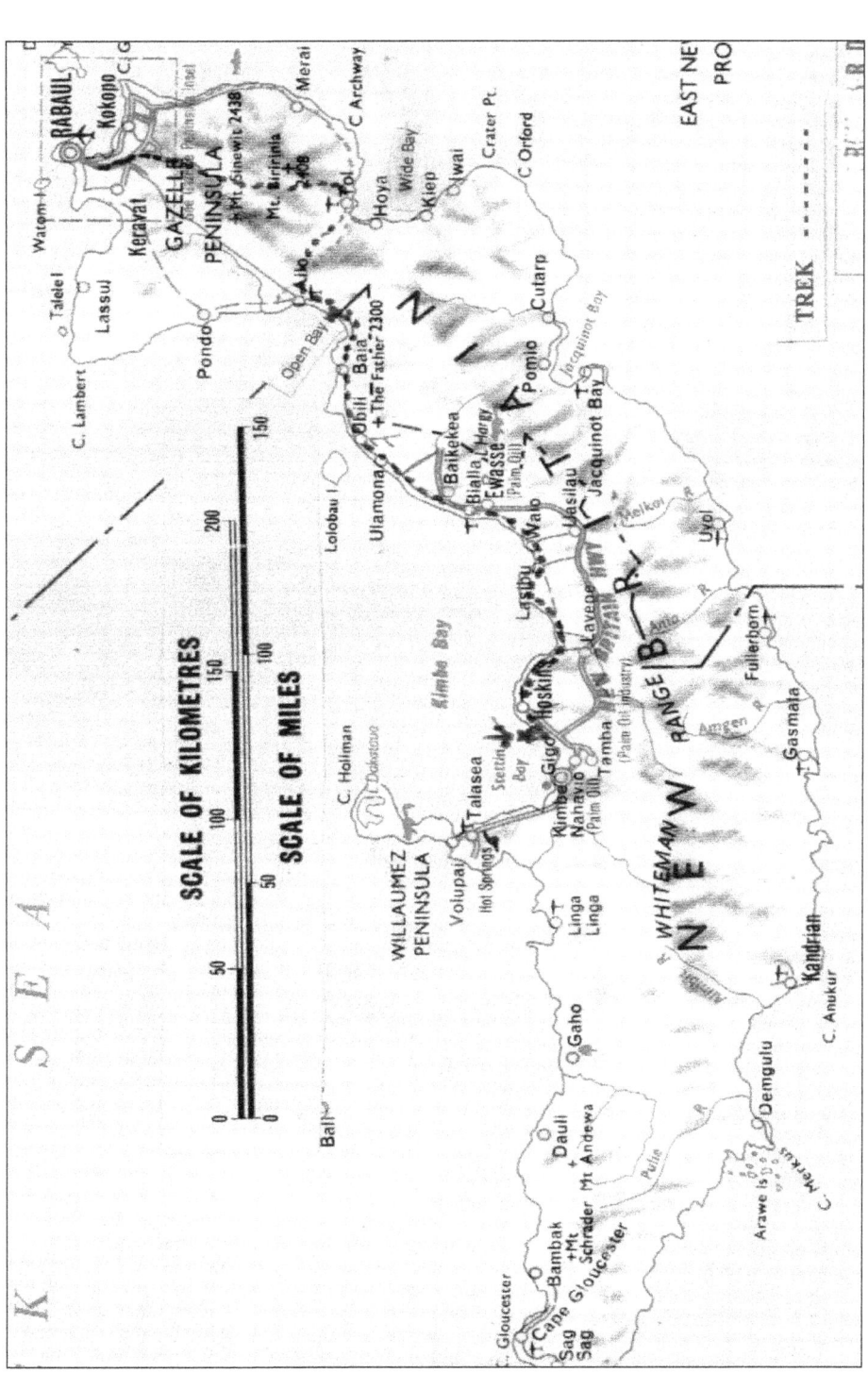

THE BEGINNING

He was ten years of age.

Wide-eyed with amazement he wandered amongst the crowds of people from every part of Australia and the Pacific Islands. They had gathered together to celebrate a great religious feast at the Eucharistic Congress in Newcastle in 1938.

His eyes were fixed on a couple of thatched bamboo huts erected in the pavilion of the showgrounds and he caught the eye of a handsome fuzzy-haired young man, clad in a colourful lap-lap, a big hibiscus flower in his hair, and a large white-toothed smile beaming from his ink black face.

"Please come and enjoy that which we have to show you of our homeland my friend."

This was not what Tal had been taught at school. The island natives were supposed to be a backward lot, suitable only for work on coconut plantations. They may be cannibals. Most certainly they were only just civilised, yet here was a voice so beautifully soft and friendly, using the English language with such precision and assurance.

"You appear surprised my friend. Do you wish to converse with me? I would be most interested in learning something of your lifestyle and ambitions. I am confident we share something which could be beneficial to each of us."

The boy shook off his feelings of astonishment and began to warm up to this so-called savage who obviously wanted to become a friend and who was capable of using the Australian language with such excellence.

"What's your name mate? My name's Tal."

"My mother calls me Bel Bel; not my official name but dear to her. It represents the song of many bells at our Mission near Rabaul in New Britain. Where is your home Tal?" he asked.

After this breaking of the ice, the two spent what was left of the afternoon exchanging experiences, names, addresses and promises to correspond on a regular basis. Tal was particularly overcome with the fact that this tall good-looking boy, who was the height and size of most of the 14- or 15-year olds he knew, was incredibly the same age as himself.

"What the devil do you eat to make you so big mate?" brought a hearty laugh.

"We mature much more rapidly than white folk in as much as physical structure is concerned, but please do not allow that to influence you. I believe I have discovered a mutual soul in you, and it is my honest wish to be your friend."

It was time for Tal to rejoin his school group and to undertake the sixty-mile journey back home. The two bid each other farewell and it was to be four years before they would meet again.

Tal spent those next four years enjoying the richness of life in his country town in the Hunter Valley, where every school holiday was spent on his grandmother's farm, nestled in a valley surrounded on three sides by majestic timber-clad mountains. He loved the lovely sandy creek, where hours every day were spent enjoying life to the full; splashing around in noisy play with his brothers, or taking long walks in the bush, inhabited by kangaroos, dingoes, wombats and rabbits. He developed high skills in trapping the latter and on occasions, accompanied by twenty-year-old Uncle Jack, was allowed to use a rifle to obtain food for the farm dogs and to bring home a 'roo tail for soup, a most appreciated delicacy in those days.

He enjoyed rising early to sit with Jack drinking tea, seated on wooden blocks right beside the enormous fireplace glowing with fire. There Granny cooked all types of delicious foods in big camp ovens hung on an array of hooks suspended above the coals. It was a magic place of tall tales and contentment.

At milking time, he would go down to the dairy with the frost cracking the grass underfoot and nestle against the warm bellies of fat dairy cows as he helped with the milking, done by hand.

It was a happy time for Tal, this time of learning about life, steadily growing, admiring his big dad who worked every day of the week fighting against poverty as he did his best to keep his family of five boisterous and hungry children during the big depression of the thirties.

He was taught the real values of his life; respect for his parents and for all people including himself.

"Give respect to everyone and you'll receive it in return," was his dad's favourite advice. He was a proud man who would not accept any form of charity in those dark days but taught the necessity of looking after yourself by working at your best level and to help others where possible.

Tal loved his dad and was truly mortified at being struck by him, which occurred, just once for a trivial offence. After sharpening a pencil with Dad's cut-throat razor and noticing the razor had lost its edge and gained a couple of gaps, Tal decided he'd better do the right thing and resharpen it. Alas, he failed to flip the blade over the leather-sharpening strop and cut the thing in two. One half was used to administer a single resounding whack where he least needed it. This was delivered in the company of the whole family and Dad's words, "This hurts me as much as it hurts you" seemed to Tal a bit of a fib.

School days and years hurried by, and by the time Tal was thirteen and a half years of age he had spent three and a half years in sixth grade. Although studying was a breeze and he had gained a bursary in each of these three years, which qualified him for college, this had been denied him because of family finances in those difficult years following the Great Depression. He realized he was now wasting his time in the country primary school and decided to go out into the world to work.

For three years he toiled happily at various ventures and gained valuable experience in practicality and social confidence.

During all of this time, he exchanged correspondence with his native friend and eagerly awaited the compelling reading in each of his letters. Tal had religiously kept every letter and would read them over and over, promising himself, "One day I'll get up there and see this wonderful island." He felt he already knew it backwards as Bel's letters were of such an explicit nature. The romance of a tropical island, swaying palms, flowers, the people and their customs and the adventure of seeing, feeling and hearing a real active volcano was truly something that could not possibly be ignored.

ADVENTURE

The year of '41 was almost at an end. A letter arrived from Bel informing Tal that he had obtained a working berth on a Burns Philp trading vessel and expected to be in Brisbane from Christmas to New Year. "Is it at all possible for two great friends such as ourselves to meet again after all these years? I pray that you have no special commitments which would prevent you from coming. It would be a tremendous disappointment to me. The first thing I shall do on my arrival shall be to look for a letter from you at B.P. office, hopefully to confirm my greatest wish at this time."

Tal was completely overcome with excitement. "Could I go Mum? What do you think Dad? I've got twenty quid in my bank, plenty to see me there and back. You must admit that after being pen pals for so long it is really important to me."

His mum was obviously against such a big step, having led a very sheltered and religious life.

"You're still only a boy and small for your age. How can you be expected to take care of yourself against the type of people you'd meet in a city? I bet you'd even miss out on going to Mass on Sunday."

"Mum, you know that Bel has been brought up in the church," replied Tal. "His dad has been with the Mission nearly all his life. How can we miss out? Come on Dad, you know I'd be alright, don't you?"

"Eily, let the boy go," spoke his dad. "He's been responsible so far and I see no harm. He has to grow up someday. A trip like this will do him the world of good. It will be lovely to hear his stories on his return."

After agreeing to abide by many rules and assuring both parents of his good intentions, he was finally granted permission.

The next week seemed interminable, but finally, he was on the train. His dad was the last to speak to him, "Be good son! Here's a bit more to help out. Don't spend it all at once, but have a good time."

Tal objected to taking the nine pounds thrust into his shirt pocket. "Dad, that's a lot of money. That's nearly a month's wages for you. I really have enough." However, his dad insisted with a final, "Don't tell Mum," and with a big grin, stepped down, and Tal was on his way.

The trip seemed never-ending. He could not bring himself to sleep, but watched as the countryside skidded by and marvelled at the changes of scenery as he was carried slowly at times, but very, very surely towards his destination. His mind was awhirl with thoughts. What do I say to him? Will I know him? Will he know me? These thoughts finally took him to sleep, to be wakened with a great noise and bustle.

"End of the line, everybody out. South Brisbane." Tal was up like a shot, grabbed his port and leaped from the train for fear it might take off and carry him to lord knows where.

He found himself mixed up in the biggest, noisiest, most frustratingly hurrying crowd he'd ever seen. He said to himself, "I'll sit here till the whole lot have gone. They've got to be mad."

Soldiers and sailors were everywhere he looked. They all seemed enormous to him as they tramped by lugging great round kit bags on their shoulders.

A group of seedy looking characters staring at him forced Tal to stop a grey-haired, kindly-looking railway porter and seek information.

"Come with me son. Sit in the office for half an hour and I'll see you safely on your way. I'll be off then and I'll take you down to the docks myself."

After an incredibly exciting ride in the sidecar of an ancient motorcycle, through blaring traffic, skidding around corners, taking, it seemed to Tal, the most amazing risks, they reached the sights, sounds and smells of the Brisbane River. "There's the place you're after, the Burns Philp wharf office. Hope you find your mate son." Expressing his gratitude, Tal grabbed his port and headed towards his goal.

THE MEETING

The wharf was a busy place. Trucks were moving this way and that way. Men were wheeling carts of goods out to the edges and waiting for cranes to haul them on board the rusty-looking old ships. Men, obviously crews of these ships, stood around in groups engrossed in very serious looking conversations. Tal, of course, being oblivious to their concerns and unaware of the dangers faced by these brave souls, set about the business he'd embarked upon.

He was about to enter the office with his prepared questions in mind, when he spied, standing by the doorway, two very big, husky, black-as-ink natives, dressed in spotless white shirts, equally spotless wrap-arounds (skirts thought Tal), bare feet, and quizzical looks on their faces.

One of them stepped towards Tal. "Masta, would you give me your name please."

Tal stood almost dumbfounded. He knew the face of this big man. It couldn't be Bel. It must be his big brother. "My name's Tal, what's yours?"

"I am Bel Bel of Rabaul."

The two stood then in silence. As they studied each other, neither could hide the emotion of the moment. Speech refused to come to either for a full minute and both were to blink back the beginnings of a tear in their pleasure and embarrassment.

A great shout of laughter bursting from the other native, came just in time to bring both back to earth. "Tal, you have shrunk. What have you been eating?"

"Bel, I'll have to find some of your tucker to try and catch up with you. You look to be twenty-five years old, at least, yet we're both the same age. I find it hard to believe it's you." Tal was just five feet and one inch. Bel was almost six feet.

"We must not worry over such things Tal. The measure of a man is in his soul and in his mind. You have proven to me through your letters that you are honest and without prejudice. Not very many white men have displayed these feelings toward us. We must now go to a quiet place where we may renew acquaintance personally."

"Bel," said Tal, "if you call me 'Masta' again I'll have a go at you, big as you are, besides, I've been on trains for about thirty-six hours, I'm filthy with coal soot, I'm very tired and I'm hungry. I need a bath."

Bel, with his thorough command of the English language, made Tal feel a little uneasy, but he felt the genuine friendship and was content. Maybe he'd teach Bel a bit of the Aussie language while he was here. Still, it's hard to teach an old dog new tricks, but he was not going to let Bel trick him into using his way.

At that moment, a jovial booming voice roared from the deck of the ship. "So, you've found your mate Bel Bel! Fetch him aboard! We'll have a cup of tea and see what we can do to help him."

"Thank you, Sir. That is particularly kind of you. If you would allow him to bathe and change, he is welcome in our quarters," said Bel.

Tal was ushered up the gangway, along the deck and into a sparkling clean cabin, an enormous contrast to the outside of the ship, which had definitely seen better days.

After showering and climbing into clean clothing, he joined the others, and during tea and cake, made the acquaintance of his hosts.

Bel's friend, though not as familiar with the Australian language, proved to have a delightful sense of humour. Though a trifle reticent and shy in the presence of white men, he joined in the animated chatter. He made one serious comment. Referring to his ship's officer and to Tal, he made his observation, "You two fella, number one Masta." His name was 'Huli' and his officer was often heard to refer to him as 'hooligan', much to his delight.

The officer, named Cliff, a genuine Aussie seaman, was highly respected and accepted the title 'Masta Cliff' from both Bel and Huli.

"You may as well stay on board for the next few days Tal, a lot safer than traipsing around Brisbane. Some nasty characters around here would skin you very smartly. The boys will get you settled in and you can all go out tomorrow and see the sights. I'll leave you now to earn my keep."

Tal was as delighted as were Bel and Huli and the next few hours were spent telling each other stories of home, comparing their few photos and planning for the 'morrow.

An hour's fishing produced two fairly large mulloway from alongside the ship, which were quickly cleaned and transported to the galley with great pride. Huli turned out to be ship's cook and promised, "Number one kai-kai" for the evening meal.

"If you would excuse us Tal, we have some duties to perform for a short while. We shall return as soon as possible. You may rest here or take a walk. I would suggest you leave any valuables with Masta Cliff as there are people hereabouts who will gladly relieve you of your money," advised Bel.

Tal couldn't believe his good fortune in receiving such hospitality and happily wandered along the wharf and surrounds for an hour or so.

On his return to the ship, he was pleasantly surprised to see his travelling clothes washed and hung to dry on the rear deck. In the cabin, he found his pyjamas neatly folded and laid out on one of the bunks and his empty port beneath a cupboard.

He decided to lie down for a short spell and quickly drifted off into deep sleep.

Bel wakened him at dusk and they joined the others for the evening meal. Huli had excelled himself with the kai-kai and was congratulated by all, to his great delight.

The conversation was of the ship and the sea and the islands and Tal's mind was fairly awash with admiration for these men, black and white who dared the elements of which he had no experience.

Plans were made for the 'morrow and it was not long before the conversation began to wane. There were yawns all round and it was time to retire for the night.

With his two friends, Tal slept to the gentle occasional bump of the ship against the piles as other craft glided by in the night.

In what seemed like seconds, he was awake and it was dawn. A great day lay ahead.

KNOWLEDGE GAINED

To Bel and Tal, the tram trip to the city was 'old hat', both having experienced the rocking, rattling, jerking ride some six years previously. To Huli, however, it was just the greatest thing that had ever happened. At first nervous and apprehensive as the tram approached, he gathered his courage and clambered aboard with his two friends. He hung on grimly at first and then began to enjoy himself. His serious countenance changed to a smile and he then startled every soul in the tram with a great laughing howl of exhilaration. "Py God Bel, tis pfella go faster all phellas bugger up pfinis quick time."

The city centre too was a grand experience for the three friends and by the time they returned to the ship in time for Huli to cook for four hungry mouths, exhaustion was the order of the day. Tal was not allowed to assist with the washing up despite his offers to help. "Galley belong me Tal," was Huli's proud boast.

Cliff, who was forced to spend all of his time onboard while others of the crew were granted shore leave, took Tal aside. "Tal, I don't think it's a great idea for them to become used to calling you by your first name only. If Tal is prefixed by Master it is quite acceptable. It's a long tradition and a deviation would be quite unacceptable in the islands."

"I'm sorry to have to argue on that score Cliff," said Tal, "but I've been reared to believe that all men are equal. I have no master, but God. I believe they have the same rights as myself and wouldn't dare to place myself above them. Please don't take offence."

"A highly commendable attitude my boy, but you shall have to find out for yourself that this is the way it is up there," replied Cliff. "We are expected to preserve the status quo but still to treat them fairly and honestly to retain their respect."

Tal did not spend any more time thinking about this conversation. To treat either Bel or Huli as anything but equals was truly preposterous.

While sitting on their bunks a little later, Bel shamefacedly admitted, "I'm sorry to tell you that I overheard your discussion with Masta Cliff. He is a good man, but as he said, he is only following tradition that has been with us for a long, long time. To us, the word 'Masta' only means boss and we are expected to treat all white men as such. It would not be a wise move for someone such as myself to speak out in this regard, but Tal, I am deeply moved to know how strongly you feel for us, having known us for such a short time."

Next morning, after completing minor chores, Tal and Bel readied themselves for another foray into town. As they were doing so, Masta Cliff appeared and rolling a cigarette, made himself comfortable. "Thought I'd drop in for a bit of a yarn before you go boys. The skipper will be coming back aboard tonight so be back early in case we need you. Have a good day in town and keep away from those floozies in the Valley. Don't take all your money with you Tal, or you may lose the lot. I'm giving you your money early Bel as you may not get another trip into town." He produced a ten-shilling note and handed it to Bel who appeared extremely pleased to receive it.

"Now my good friend," he said as they tramped down the gangway, "I shall be able to purchase gifts for my parents. They will have missed me for Christmas, but now I am able to please them on my return."

"Gee, ten bob isn't much for a week's work Bel. They're a bit stingy, aren't they?"

"If I were fortunate enough to receive 'ten bob', as you put it, for a week Tal, I should become wealthy in no time at all. The remuneration paid to me and my fellows is 'ten bob' a month and my keep."

Tal was shocked beyond words. At his last place of employment, he had been paid thirty-five bob per week, ten of which to be the going rate for bed and board. He resolved to see that his friend would not go home without something worthwhile.

"Bel, I wasn't here to see you at Christmas time, and I'd like to give you a present now. We've been good friends for a long time and I'd like you to have this." As he spoke, he handed over two one-pound notes.

Bel was overcome. "This is more money than I have ever held in my hand at one time. Forty whole shillings. That is four months' pay. Tal this is too much."

Despite all of Bel's determined protests, Tal insisted and won the argument by inferring that a refusal would be an insult.

Bel replied, "I am extremely grateful for such a generous gift. Would you be insulted if I were to share my good fortune with my friend Huli, whom I have known and respected since childhood?"

Considering the fact that he still had around twenty-five pounds, Tal explained that he intended to do something for Huli and decided he would give him one pound and that Bel could keep his Christmas present to do with as he wished. This naturally caused Bel a great deal of excitement. "Would you mind waiting for a few minutes, Tal? I must inform Huli of this wonderful gesture of yours." With this, he galloped up the gangway with his lap-lap clutched high to gain more speed. This caused a great deal of merriment to the wharf labourers who were sitting around having smoko. Many were the crude jokes referring to the type of, or lack of, underwear worn by Scotchmen and 'boongs'.

Tal, having been reared in a strict religious atmosphere reddened with embarrassment, which only served to stimulate this raucous crowd. He was genuinely relieved when Bel and Huli arrived back on the scene. Their size and stern glares silenced the motley crowd of comedians at once.

The happy trio moved off, the two New Guinea boys chattering in their own language as they walked up into the Fortitude Valley conglomeration of trams, traffic and crowds. Bursts of laughter from them continuously elicited a "What did he say?" from Tal.

A good inspection of the various stores came next, followed by lunch, and then began the serious business of disposing of their new-found wealth in the form of gifts for their families by Bel and Huli. Tal also purchased ten yards of beautiful dress material, two excellent knives (one for himself), and a set of two pipes (as Bel had informed him that his father indulged in an occasional smoke.) "These are gifts from me to your parents Bel. I hope they like them."

"They will be absolutely overcome with gratitude, Tal," he replied.

Soon, weary with walking and shopping, they agreed that the job was done, and struggled with their arms full of booty back to the wharves and the ship. On their arrival, they found the captain, Mr Ward, had already arrived.

"Well young man, I hear you've taken over my ship and my crew. My mate has informed me that you're a good lad and have paid your way doing the odd chore. You're welcome to stay on board until we leave in a couple of days." He was not a large person. His voice was soft and friendly, quite the opposite of that envisioned by Tal, who expected all captains to be large swashbuckling bullies with stentorian tones, shouting continuous orders. He thanked the captain for his hospitality and assured him he would not forget such generosity.

He then retired to the cabin with his two friends, who stowed their treasures and made themselves quickly available to carry out any tasks allotted to them.

Huli took off to his galley, Bel went off with Masta Cliff and Tal stood by the rail watching the workers and wishing he could one day take on employment in such a convivial atmosphere.

After the evening meal, sitting on deck with Bel, Tal asked every question that came to mind regarding life at sea and indicated his interest. "Mind you mate, I've only seen the ocean once in my life. That was six years ago when I met you in Newcastle. Do you think I could handle a job on a ship?"

"Tal, this is the first voyage I've ever made. Huli persuaded the captain to enlist me on this voyage for the sole purpose of meeting you here. The ship will be going to Rabaul and returning immediately to Brisbane. I shall ask Masta Cliff to speak to the captain. Maybe he would allow you a working passage for the couple of weeks it would take. I could show you Rabaul, you could meet my parents and friends, and I'm certain you would enjoy yourself immensely."

Tal drifted off to sleep that night agog with anticipation and some nervous perceptions of the sea and its dangers.

The next morning (31st December 1941), at breakfast, Cliff sat back, lit up a cigarette and casually announced, "Tomorrow morning we set sail for Rabaul," and proceeded to instruct the crew who were now back on board regarding their various duties. When this was completed, he directed his attention to Tal. "Well young man, so you'd like a taste of the sea eh! I'll say this for you. You're fairly game or fairly foolish to want to go out there considering the way things are with the Japanese. If you were my boy, I'd send you packing for home. Anyway, the skipper will see you in his cabin in half an hour; you need not expect too much help from him, believe me."

At the appointed time, Tal presented himself. "Good morning Mr Ward. You asked to see me."

"I've heard your request from two sources and I'm not sure that I should help you. These are dangerous times and I do not wish to take responsibility if anything should go wrong."

"Please Mr Ward, I'll work for no pay for the return trip and I'll do any job you ask of me. This is a chance of a lifetime for me and it will give me another week with my mate, Bel. Would you ring Dad's place of work so that I could get his permission? If he says no, then I'll stop being a nuisance."

After Tal had given him the necessary information, the captain was connected, told Tal's father of the request, then handed Tal the phone. "Good day Tal. I've heard your plan and won't stand in your way, but don't let your mum know about it until you come back. You know how worried she gets. Write her a letter, but make any other excuse you like to cover your absence. Look after yourself and work hard for the privilege of such a great favour."

After thanking his father and making his promises, he ended the conversation. Captain Ward then became very stern as he advised Tal regarding the ships' rules and acceptance of discipline, ending with "You will always refer to myself as 'Sir'. The same applies to the Mate. You are a type of guest, but the rules apply. Good luck and enjoy your stay with us."

Tal was overjoyed and raced to inform Bel of his good fortune. The two shook hands and laughingly thumped each other with great glee.

On the approach of Masta Cliff they both shouted the good news. The Mate, though not overly pleased, wished them the best and promptly began issuing orders to get back to work. "Yes Sir," piped Tal in his excitement. The first time he'd ever had occasion to use this phrase.

"You may at times refer to me as 'Mr Cliff' Tal in order to maintain rank in its proper perspective, but I feel we're going to get along fine."

Tal's letter to his mother informed her that there were people in Queensland who lived so far outback that they only came to town every two or three months for supplies. He explained that in order to see more of life, he had decided to take a working trip, not to worry if she didn't hear from him for some time, as he'd be in good hands.

He decided he'd not told any lies, just didn't offer the whole story. He knew his dad would back him up and it would only be for a couple of weeks anyway. This necessary item out of his mind, he placed the letter in the mailbox on the wharf and became lost in the excitement ahead.

The day progressed slowly for Tal, but dusk saw the ship fully loaded. Perishables like great sides of frozen beef and mutton, vegetables, butter, etc., were the last items to come aboard and to be packed in the cavernous freezing room down in the bowels of the ship.

As an early departure was ordered every man was aboard, some complaining about missing out on the New Year celebrations going on in the city.

VOYAGE TO RABAUL

Even the noise of these celebrations till well after midnight could not be blamed for Tal's lack of sleep, as the excitement was proving too great for him to miss a moment of this monster adventure. It was still dark when 'all hands' were called and the ship left her ties with the shore. Shepherded by a tug she slipped along the river, out through the long rows of beacons and turned north into the deep channels of Moreton Bay.

The engines increased to a powerful throbbing, the ship lifted her bow and charged past the tug with a farewell blast of her horn.

Dawn saw Caloundra slipping by astern and the ship began a gentle pitch and roll, which caused Tal to forego his breakfast. At his refusal to eat, Masta Cliff joked, "Good to see you don't believe in wasting good food, but that's good, you'll be hungry at lunchtime as you get used to the feel of it. We've all gone through it and don't let anyone tell you he hasn't."

Chores to be done, though regular, were fewer than Tal imagined and he was able to enjoy standing against the rails right at the bow. He never tired of watching the powerful bow ploughing its way through the blue water, seeing the sun above shooting rays of light to a point way down into the blue depths. Occasionally, a school of dolphins rode the bow wave for hours and he marvelled at the sight of flying fish. In their fright to escape this monstrous hulk invading their territory, they would shoot out of the water, spread their long wings and glide for amazing distances. On beginning to lose speed, they would flick their tails rapidly on the surface and regain flying speed for another long glide.

As the ship was plying inside the reef, he saw turtles galore, each making a sudden dive at their approach. The water was crystal clear and they could be seen hurrying down to great depths. Around the stern of the ship beside the wake, he saw at times, huge sharks cruising effortlessly and keeping up with the ship with apparent ease. "Hate to fall overboard Bel," he remarked.

In every direction he looked, Tal saw only water, and being used to the bush, he began to long for the sight of hills and trees, somewhere to go for a walk.

During the days that followed, he saw trees growing out of the ocean. It was explained that beneath them were tiny coral atolls, which could not be seen because of distance. He saw a long black cloud sitting atop twelve black columns out on the horizon. These were waterspouts and quite common. On one occasion, while leaning over the bow rail he looked ahead, and in the perfectly calm water there stretched for miles, a seething white barrier approximately two metres in height.

He called loudly and was joined by Bel, who also became exceedingly alarmed. Despite their frantic calls to the bridge, the ship ploughed relentlessly forward to certain destruction on an obviously shallow reef.

The ship, however, struck this phenomenon with a soft 'boof', lifted a little and settled down to forge ahead without harm.

Our two heroes drifted back to the foredeck, vacated in great haste, to be told amidst a deal of laughter, that the apparent danger was simply the meeting of two ocean currents caused by tides retreating through from the Arafura Sea and meeting the currents headed down the east coast of Australia.

Another eye-opener for Tal was seen well out into the ocean. There had been no land in sight for a whole two days, and then a small speck on the horizon dead ahead caused some excitement from the bridge and from the crew. As they drew nearer, the object turned out to be two large log canoes, 'lakatois' joined together by lengths of bamboo. On this structure was a platform, also of bamboo, lashed tightly in place. A single mast stood high in the front of this vessel, about twenty feet in height and on the platform sat at least a dozen natives, including children. Mats covered fruit and vegetables. Two pigs plus a few fowls were tethered upfront. There was not a breath of wind to help them on their way, but they all waved happily as the ship cruised past at a very slow speed. It was later explained by the ship's mate that this was a common sight and had been going on for centuries. Long distances were travelled from island to island, the navigation incredibly accurate without the assistance of compass, charts or any means other than the position of the sun and stars.

Friday, 9th January. At about dawn, the small ship made its way into Rabaul Harbour. All that Bel had told Tal came true, as he saw five volcanoes in one quick sweep of vision. As if to greet the ship, 'Matupi', the smallest and youngest, gave a rumble not unlike thunder, and belched forth, into the still air, the most enormous smoke ring imaginable.

With no breeze in the still air, there hung, over the harbour, a thick smell of sulphur, particularly offensive to the nose. Bel assured Tal that this was a normal occurrence in Rabaul and even though he and his family had been through two great eruptions, they had survived and took no notice when the earth heaved and rumbled.

"Mate, if I lived here, I'd go live somewhere else. It looks flaming dangerous to me."

At this Bel laughed, "If I can get used to it Tal, then so can you. Anyway, you'll only be here for a few days so don't worry."

As the ship berthed and unloading began, Tal noted without surprise a number of army officers, all looking very serious and self-important, coming aboard and being escorted to the captain's quarters. He placed no importance on this occurrence, as he knew the ship was carrying supplies destined for the army.

Masta Cliff put in an appearance and told Bel and Huli they were free to go and "thanks a lot boys, you've been a big help." To Tal he advised, "Leave your gear onboard mate, and keep in touch as frequently as possible. We don't know how long we'll be here, but at least until we've unloaded. Don't go too far away now."

Tal assured him he'd do so and hurried off to join Bel and Huli. Huli lived down on the southern end of the harbour and after his goodbyes, set off to join his family.

Bel was luckier. A fuzzy-haired man clad in a clean white lap-lap accompanied by a woman in her 'mother hubbard' (a lap-lap and a large coloured loose blouse) descended on Bel with loud cries of excitement and obvious love.

"This is my mother, and my father Joseph, of whom I have often spoken to you. Mother, Father, this is Tal."

Tal was overwhelmed by their greeting. He thought Bel's dad would never stop shaking his hand. His mother was shy, but "very, very pleased to meet you Masta."

"Bel, will you put them right about this 'Masta' business before we go anywhere else."

A long conversation took place in their own language, as Bel's mother was not as fluent as her husband and son in English. Eventually, they agreed that in private they would use Christian names.

They immediately insisted that they all retire to the family home, which was about a twenty-minute walk up into the jungle. Bel firstly asked them to wait while he and Tal went back on board to retrieve Bel's meagre kit plus the gifts they'd purchased in Brisbane.

On leaving the ship, Tal was accosted by an army officer complete with a large pistol at his side and a red band on his upper sleeve. Impatiently, Tal answered his questions and was told to make daily contact. He hurriedly agreed and made haste to join Bel and his family.

Tal, despite his protestations, was not allowed to carry a single article. "You are our honoured guest. We have read all of your letters and still have every one of them," stated Bel's dad.

The journey through the bush was a delight to Tal. It was not only very beautiful but quite a circus as well. They met, in ones and twos, at least fifteen people coming down towards town. They stopped each time, were given a rousing run down on the situation, turned around and accompanied the four of them to the village. The village was as Bel had described it with very clean grounds, swept clear of leaves. The well-spaced homes were built of bamboo framework, thatched with long grasses and lined with woven mats of coconut fronds. Flowers, ferns and vines surrounded most of the homes. There seemed to be chooks everywhere, plus a half dozen or so fat black pigs.

Big wide eyes of children surveyed every move he made and his smiles were returned brilliantly in their shiny black little faces.

A large covered verandah fronted Bel's home. Fresh coconut milk was served and the village people were treated to a complete run-down on every single detail of Bel's trip and his meeting of this one, 'Tal' after so long. The gifts were then given to his parents, who were speechless with pleasure.

Before the crowd departed, something occurred that brought tears to Tal's eyes and gave him a memory that would stay with him for the rest of his life. One by one, then altogether, they sang with great gusto and much feeling, his mother's favourite hymn, 'Faith of Our Fathers'.

"That particular hymn strikes a deep feeling within all of us who share our religious convictions. I saw the emotion it created in you and it makes me happy for you and proud of my people," stated Bel's dad. "Everyone in my village knows of your long correspondence with my son, and of the fine mutual feelings you share regarding life. I look upon you as a second son. I am beginning to overflow. I thank you most sincerely for your gifts. The pipes will be a great boon to me. Father John at the Mission smokes a pipe when he is thinking out a problem and he is a very wise man. He looks a wise man with that pipe. Maybe now I also will look a wise man. The people will surely look up to a man who smokes such fine pipes as these."

Tal saw the twinkle in his eyes and realised he was being led along. A good laugh was enjoyed by all and the pipes were passed around to be admired and tested as a token of wisdom by all and sundry.

At Bel's insistence, he made an inspection of the home, which revealed spartan furnishings. Each of three bedrooms consisted of walls of woven fronds and a soft woven mat on the floor overhung by a clean white mosquito net. The living area was larger containing a table about a foot high, surrounded by cushions. Every wall was hung with the odd religious print and photographs of groups of people near a small neat church. There were also photographs of the small family shown with priests, nuns and the occasional bishop.

The kitchen was an attached room. The roof was almost conical in shape with a large hole in the centre. A square of corrugated iron was suspended above this opening in order to prevent rain from entering the room. The bamboo rafters were black with smoke, which would billow up from an open fire on a slab of concrete about four feet by four feet. Hooks hung down to suspend cooking utensils and curtains covered knives, forks, spoons, etc., hanging on the wall.

Bel's mother was ecstatic when Tal announced, "This is almost the same as my grandmother's kitchen."

The whole family insisted that Tal stay for lunch and were pleased to receive his grateful acceptance.

The meal consisted of cold chicken and rice covered with gravy followed by a salad of banana, pawpaw and five corners, a fruit not unlike plum in taste. The prepared food had been taken from what was called a 'cool safe', a wooden framework, square in shape, with the top smaller than the bottom. This was covered in hessian. A tray of water on top with strips of hessian leading the water down and saturating the sides kept food almost chilled. Tal was familiar with this practice as his grandmother also used this method.

The meal was preceded by the saying of grace, giving thanks for the safe return of their son and calling for blessings upon their new-found friend.

Joseph related to the two boys that an air raid had occurred a few days earlier by Japanese aircraft. "They made a lot of noise but did not manage to do much damage. I think by all accounts that the Americans may chase them back home soon."

Tal felt no concern at the time. Little did he know that the near future would change his life forever.

The sky darkened suddenly and just as suddenly, the rain crashed down. It did not pour as Tal had ever seen it. There was no other word for it, but 'crashed'.

Tal felt sure that this bamboo house could never stand up to such an avalanche of water. A low rumble grew into a great roar in the earth and this proved to be his first horrifying experience of an earthquake. He had to shout to be heard, and on top of it all, everything began to heave and shake as though some giant hand was endeavouring to pull the home from its foundations. His fear was indeed great, and did not diminish until the shaking was over and Joseph shouted in his ear, "Welcome to Rabaul!"

A little later as the torrent eased to a heavy downpour, he learned that the rain, though late on this day, was an everyday occurrence and that these earth tremors could be felt sometimes twenty to forty times a day. This particular one, according to Bel, was about as bad as could be expected, but most caused no anxiety.

The rain was not expected to ease until about midnight, so Tal readily accepted the family's invitation to spend the night with them. He relished the evening meal and was warned to eat only a limited amount of fresh pork. Bel's mother explained the dire consequences to be expected if he ate too much. "You do plenty running." She was a handsome, kind woman, but had not received the standard of education enjoyed by her husband and son. Her halting English was a source of much amusement to Tal. When speaking to her family in their own language, however, she was able to outdo both Joseph and Bel, who were lucky to get a word in.

A cold shower was in an enclosure under the tank at the end of the kitchen. A clean lap-lap served as pyjamas, and it was time to retire for the night. The woven mats proved to be unusually comfortable and Tal had no difficulty in drifting off to sleep. In these strange surroundings, he felt safe and comfortable and was warmed by the genuine, frank nature of his new-found friends.

Next morning dawned clear of rain. The surrounding jungle was alive with the sounds of birdlife. There were pretty little songsters busy among the flower gardens and parrots high up in the foliage raising their raucous voices.

The village fowls were scratching about and the whole picture appeared to Tal as something that belonged in storybooks.

Bel joined him on the verandah. "It's so good being home again Tal. I much prefer this atmosphere than the sea or the city, even though it all proved most beneficial to me. I suppose we all think at times that everyone else is enjoying life more in faraway places. I belong here and here I shall stay, maybe to fulfil my parents' wishes and study for the brotherhood or even priesthood. I am plagued with doubts about this though, as I have my eye on a certain young lady who would make an excellent wife for any man. Maybe I am not sufficiently strong to endure celibacy. Time will tell as my father says."

"Well mate," replied Tal, "we have both been given a very full appreciation of right and wrong. I intend never to do the wrong thing, and to follow the rules given to me over the years, but I feel deep down that I should find my own way without guidance. They say life is short, so I intend to see and experience as much as possible before I settle down. What do you say we go to town, check in at the ship and do a spot of shopping? The little I've seen of Rabaul makes we want to see more."

Bel heartily agreed and after a breakfast of toast, liberally heaped with warmed up leftovers from the previous night, topped off with tea, the pair disengaged themselves from their reluctant hosts and proceeded to town.

The walk to town was a delight to Tal. The high walls of the jungle were truly impressive. A further delight was provided by a horde of a dozen or more shiny black, fuzzy-haired village children in their best colourful lap-laps, either following along behind, or walking backwards in front, laughing and chattering for the whole distance down the long ridge that led to town. Tal joined this happy crowd in attending Mass as he'd promised before his departure from home.

On reaching the wharf, Bel elected to stay with the children while Tal proceeded on board to an unexpected and decidedly unfriendly reception.

The captain and Cliff promptly ordered him to the captain's cabin where they began berating him furiously.

"Firstly," began the captain, "where the hell have you been all night?"

"Sir," said Tal, "you told me I was free 'till the ship unloaded and here I am checking in. What are you so angry about? The ship is still being unloaded. I spent the night with Bel and his family because it rained so hard and I had no raincoat."

Cliff joined in with a long harangue, "Captain Ward told your dad he'd look out for you and I myself warned you about getting too thick with these people. It's only inviting trouble if they get too familiar. Give 'em an inch and they'll take a mile. Keep it up and you'll start to smell like one. We're the masters and they the servants. They know their place and we know ours. Don't mix with them anymore. Do you hear me?"

Tal's anger built up during this beration and he exploded, "Captain Ward, you were very kind to bring me up here. I worked for that privilege and don't owe you a penny. Those 'people' as you call them are more Christian, and in your case, Mr Cliff, more clean than you!"

"Watch it boy!" roared Mr Cliff. "I'll kick your bum clean off the ship if I hear any more talk like that!"

"You don't have to do that Mr Cliff, I'll get my gear and leave of my own accord. I'm sure I can get home on another ship."

"That won't be necessary son," spoke the captain. "We'll be out of here by tonight and we'll all be able to forget about it. You just be back on board before six o'clock and there'll be no more said."

Tal furiously left the cabin, went to the sleeping quarters, packed his port and marched off down the gangplank.

The captain and his mate were watching as he rejoined Bel. Tal gave them a mock salute, which was not acknowledged by either man.

Bel scooted the children off to play in a nearby park and then began a barrage of questions.

"What will you do Tal? How did this all come about? I'm sorry this all came about because of me, and I'm sorry they feel this way towards us. They were always good to me."

"Not really Bel. They took advantage of you as cheap labour and I could never forgive them for their attitude towards your people. I told them off good and proper and feel good for having done so. Let's forget them. They're leaving tonight, so I'll keep out of the way till tomorrow and then see if the army can get me on another ship."

Bel's spirits fell and for the rest of the day he lost his usual cheerfulness. His big cheerful grin didn't reappear until Tal reassured him that such an attitude displayed by his former employers did not dampen his deep respect for him and his family. Bel insisted that they must return to the village until other arrangements could be made.

A shopping expedition was then planned. A large box of groceries was purchased, plus a raincoat and hat, and a treat for the 'kids' and the remainder of the day was spent inspecting the town.

It was a lovely small town with plenty of shops, run mainly by Chinese. The homes were white and neat, surrounded by colourful, well-tended gardens and lawns. The streets were well laid out, and lined with avenues of trees and palms. There were parks and town gardens; the whole town looked resplendent under the backdrop of green hills, volcanoes and the deep blue waters of the harbour. Small outrigger canoes called lakatois were being paddled about everywhere, their occupants bent on seemingly mysterious business.

The skies down beyond Vulcan volcano at the south end of the harbour were beginning to cloud up, so at Bel's suggestion, it was time to retreat to the village. The procession was still within ten minutes of home when the deluge began. The raincoat was used to cover the groceries, so needless to say, everyone in the party became soaked in seconds. It was impossible to see more than a few feet, but the boggy track led them all safely back to the village.

Tal gratefully accepted a lap-lap while Bel's mother took his clothes to wash them and hang them on a line hung on the front verandah.

Bel's dad, Joseph, joined them and solemnly listened to everything that had transpired.

"I think at this moment, Tal," he observed, "that you may have acted very hastily. You may find yourself in trouble with the authorities or the army. I have great admiration regarding your ethical stand on our behalf, but please don't worry too much about us. It may not be in my time, but one day Bel will find himself to be his own master. Our people will become independent and be the arbiters of their own destinies. This I believe will come about in a peaceful fashion. You are welcome to enjoy our hospitality for as long as you need."

After lunch, with the family seated on the verandah, the peace was interrupted by the appearance through the pouring rain of a group of soldiers on their way through the village on some business or other. Tal waved. The leader called a halt, dismissed the men and to a man they tramped up the steps, shook their rain cloaks and squatted down against the walls. Most lit up cigarettes and were pleased of the rest and break from the rain.

The officer-in-charge, a lieutenant approached Tal. "You from town lad?" Tal went on to explain his position and his intention to report in as requested earlier.

"Request my foot, you may consider that an order. Evacuation of civilians is top priority now, so you get yourself down to headquarters first thing in the morning. If you don't turn up, I'll come back and personally drag you down by your ear. OK?"

Tal gave his word that he would be there and after awhile, the troops donned their gear and proceeded on their way up the long ridge that led to at least four other villages higher up in the jungle.

"There must be something of urgency occurring," observed Bel. "We don't see them up here very often except to recruit labourers from the village. They normally come two at a time, but not in such numbers. We may find out after your visit tomorrow."

On Monday, Tal reported to headquarters of the 2/22nd Battalion. The major to whom he was escorted proved to be very cordial and advised Tal that he should stay at headquarters pending evacuation. He (the major) appeared concerned that such a small and young individual should have been so far from home alone.

"I'll leave you in the charge of Sergeant Major Wilson here. He'll assign you to quarters and you'll make yourself available at all times until we're able to ship you out. Can't have you living up there on your own you know."

"I wasn't on my own Sir. I am staying with my friend. I've known him for six years and his family's been extra kind to me. They aren't kanakas. They're educated and clean and have been with the Mission all their lives."

"That may well be son, but you're our responsibility now and you must do as you are told for your own safety."

Tal had no answer to this reasoning and followed the sergeant major, who escorted him to a nearby tent occupied by three soldiers who were not present, but the evidence was there in that their kits lay on their bunks, either packed and ready to go or not as yet unpacked. He was assigned a mosquito net covered stretcher and questioned, "Where's your gear mate?"

Tal explained that it was still at the village and that he'd go back and get same.

"Out of the question I'm afraid," said the S.M. "We'll arrange to have it brought in."

Just then, Bel appeared, apprehensively requesting to speak with Tal.

"This is my friend Sir. I hope you'll allow him to come and go. He's one of the best blokes I've ever met."

"Well, boy, if you're his best friend," said the S.M. to Bel, "you'll go home and bring down his gear. Off you go and don't take all day to do it."

Bel meekly answered, "Yes Masta," and disappeared from view, leaving Tal absolutely miserable that such a fine person could be treated so badly.

He knew it would be futile to complain, as this officer exuded such authority he thought it best to remain silent. He decided he would not forgive such bad manners and would show no respect even to an army officer.

Bel reappeared about two hours later with Tal's port and accompanied by his father, Joseph.

Tal, still miserable, confided his feelings to both his friends and advised them of his fate.

"Don't take it hard, Tal," consoled Joseph. "It's always 'boy' and 'master'. We don't agree with it but it's nothing you or we can change. Just bear up with what has to be done. We hope you'll eventually return home safely as I fear this place will soon not be fit to live in. Bel will be with you as much as possible."

With that, Joseph walked away leaving the two boys to ponder the future.

There very soon came a rude interruption as a group of aircraft swooped in from over the hills, nosed down and began strafing everything in sight. Bombs fell from their wings and though they appeared small, they blew up with frightening bangs. Over they went, completed a wide turn over the harbour and repeated the performance. Crouched in a handy trench, the two friends cowered as bullets churned the earth and a nearby building disintegrated into a ball of flying debris, dust and fire.

The soldiers replied enthusiastically with ack-ack guns, machine guns and rifles without seeming to affect the aircraft. In their confidence, they sometimes flew so low that one could clearly see the helmeted faces of the pilots. Tal was completely helpless, but felt anger that he'd never known before, build up to burning hatred.

"By God, I wish I had a gun Bel. I bet I could have hit that pair. They were so close. If I only had Uncle Jack's .32."

"Tal, you've turned into a monster in a few minutes," said Bel as they crouched down to escape the onslaught.

"Well mate, you're nearly as white as I am. We're both shaking like leaves and I'll bet you'd like to have a go at them too," replied Tal.

"I'd like to have a go at the track back home first," replied Bel.

The raid lasted for fifteen to twenty minutes, the planes concentrating on one area after another and it was a very relieved, very muddy pair that crawled out after the marauders had departed.

Soon after the raid, a very worried looking Joseph came running down. He was all over Bel and shaking with apprehension. After satisfying himself that the boys were okay, he almost wept with relief. "You boys had better come home where it's safe. They're not interested in our village."

Tal explained his position, and after a while, his two friends departed, sad to be leaving him in such danger.

Early in the afternoon, the rains arrived again, blotting out everything, but making things safe from attack. Out of the downpour arrived a truck carrying a load of very drenched soldiers, all heavily armed and obviously happy to be back. Three of these rushed into the tent where Tal was forlornly thinking of all that had been happening.

"Holy smoke, what have we got here fellas. Didn't fall out of one of those planes did ya? Nope, he's not a nip. Does your mum know you're out?"

Tal suffered all of this good-natured banter until they'd settled down and dragged from him his long story.

One of the three, who seemed particularly concerned at Tal's predicament, softly told him, "I think we'll soon have you out of Rabaul and back to your mum. There'll be other ways of getting you home I'm sure. In the meantime, we'll look after you. By the way, indicating the other two, that's 'Robbie' and that other ugly bugger we call 'Dump'. One day you might find out why. My name is 'Kirk'."

Tal took to this lean, long, string-of-a-man immediately, recognising his more genteel and knowledgeable nature as compared to the others, whose garrulous talk and profanity did not appeal to him.

In the late afternoon, a bugle sounded and they all jumped to their feet and there began much rummaging and jangling as they dragged out their eating utensils.

"Got any dixies?" queried Robbie.

"He wouldn't know what dixies are, ya clot," joked Dump. "We'll find him a little bowl and spoon. He wouldn't be able to eat any more than a bloody pee wee."

"Well Pee Wee, come with us and we'll introduce you to some blokes they call cooks," said Robbie.

Kirk took over and led Tal to a long line of tents where hungry soldiers were busily tucking into their meal. His situation was explained to the cooks who produced a pair of square aluminium containers with folding handles (one fitted inside the other when empty) and a knife, fork, spoon and mug.

"What would you like for tea mate? You can have bolognaise or stew," and with that, the cook dumped a large ladle of rich stew in one dixie and some prunes and rice in the other. Tal joined Kirk who had stood in line behind him and they sat on the long benches to attack their meal with gusto. There were stares from just about everyone present and joking remarks about the size of reinforcements being sent from Aussie.

Dump announced, "This is Pee Wee and he's going to be with us till the sar major can get rid of him."

Tal bristled with annoyance at being addressed as 'Pee Wee' and told Dump so in no uncertain terms. "My name's Tal, mate, and if you persist in calling me names then I can only assume from yours that that is where you come from."

"Well said, little mate! You've got guts and a quick mind as well," spoke Kirk, "but I'm afraid that in the army, if someone is given a nickname, then he's got it for good. We've got Stinker, Stringbark, Tanglefoot, Foxey, Nosey, you name it. Actually, the men without nicknames feel left out. I'll call you Tal anyway."

Tal's ire went down with this explanation and he guessed it would do him no harm as he wouldn't be around for long anyway.

The meal over, he went to the cooks and to their astonishment, thanked them for a lovely meal. "Good on you mate, you keep the dixies. We won't skite that someone dared to thank us."

Tal followed Kirk in washing their utensils and sat in the tent listening to the men as they drank their ration of beer and pondered gravely over the war and how they might fare if the worst came to the worst.

"They won't come in here. There are plenty of places worth more to them than here. They stretch out too far and they'll break their own backs like the Krauts did in Russia. They're just pasting us for something to do while they think up something sensible."

And so it went on until the notes of the bugle rang 'lights out'. Everything went quiet and Rabaul fell asleep in the seemingly endless rain.

The next six days were spent with the army, almost every day was a terror of roaring aircraft, blasting guns and hiding in watery trenches against the fury of the Japanese. Added to this demonic existence were the irregular but unsettling frequency of earth tremors.

While walking about on one occasion, a tremor became sufficiently severe as to cause Tal to become quite unsteady on his feet. He decided to sit instead and was awed to see the earth rising and falling not unlike swells in the ocean. Coconut palms were waving, nuts were falling everywhere and the previously still waters of the harbour became choppy. Soon a blanket of pumice, shaken loose from below, covered the harbour until no water could be seen. The foul smell of sulphur on top of all this turned the place from paradise into a stinking hell. The black sand of the beaches in Blanche Bay only added to this sombre atmosphere.

Tal decided that despite his adventures, 'there's no place like home', and wished for home with all his heart.

Bel put in regular appearances, much against the wishes of his parents, who quite rightly stressed upon him the dangers involved. However, Bel was not about to abandon a friendship that had grown into a genuine bonding between himself and Tal. They were by now as close as brothers, each treating the other with the greatest of respect.

Their contrast regarding skin colour was not considered at any time by either boy, but Tal was forced to bear a few racist remarks from some of the army personnel. Often, upon defending his friendship, he risked a 'clip in the ear' or a 'kick where ya mamma never kissed ya'.

However, with big 'Kirky' at his back, he felt fairly safe from these physical humiliations and stood up worthy of someone twice his size.

"I'll say this for you Tal, you're as game as a meat ant," praised Kirk after one of these confrontations.

The army major was furious at his failure to receive any assistance from Australia in order to evacuate the civilian population. One of the ships he'd been counting on had slipped away to escape the bombing. Another was inexplicably loading copra, leaving no possibility of assistance.

Then came the reports everyone had feared. A very large fleet of Japanese ships, battleships and aircraft carriers had been seen off New Ireland and they were heading directly for Rabaul. The soldiers abandoned their madcap joviality and became deadly serious as they set about preparing to meet this frightening threat.

A group of civilians, including Tal, was summoned before army officers and informed that nothing more could be done to get them away from Rabaul and it was up to everyone to go it alone and look after themselves as best they could. Bel was with

Tal when this announcement was made. "Come with me Tal. Our people will look after you. We know our way about these mountains. You could hide forever."

"I don't know Bel. I'll let you know tomorrow after I have a yarn with the major and Kirky. I don't like the thought of running away when I could help in some way. They're all Aussies and so am I. I'm scared stiff mate, but I think they'll look after me too. The nips won't be at war with you and your people. If the worst comes to the worst, I'll get to you."

The soldiers then began moving as much as possible up into the hills, surrounding the town. Tents, cooking gear, boxes of rations, ammunition, all laboriously carried by hand with the exception of a couple of four-wheel drive 'Blitz buggies' (large trucks) that were eventually hopelessly bogged and were abandoned on the narrow track that wound up and over the hills to the northern side of the island.

Slit trenches were dug in long lines where the terrain afforded good vision down onto the beaches, which were being strung with coils of barbed wire. 'Surely no one could clamber through that lot,' thought Tal. He was ordered most emphatically not to venture anywhere near the beach as it would be heavily laced with land mines.

ENLISTMENT

Everyone knew by then that a great battle could not be avoided. The 2/22nd's three companies (A, B and C) were called together and informed of the threat that was drawing near and that fighting on a great scale was inevitable.

Tal stood nearby and listened as the brigadier urged his men to do their best and assured them of the faith in his men. He called on the chaplain to deliver a prayer. The faces of the men showed no fear and Tal was at that moment proud to be Australian. These men were facing almost certain death but showed a fortitude that caused Tal to wonder about his own performance. However, these gallant men inspired him with a new profound courage that he was absolutely certain could sustain him whatever happened.

He made a decision and after the parade, began to set about putting his plans into reality.

He spied the sergeant major and asked for permission to speak to the major, who was the company commander. "What's your problem boy?" said the S.M. "You'd better be quick. There'll be another raid soon for sure."

"Sir, seeing that I'm stuck here and living off the army, I thought I might be able to join up and be of some use," replied Tal. "I'm as fit as a fiddle and I can shoot as well as anyone here. The indication appears to be to fight for as long as possible then every man for himself. I don't want to be a hindrance to anyone. I know I'm younger than anyone else, but I'm not useless. I'd like you to state my case to the major. If he says no, then I'll take to the bush."

This discussion was interrupted by air raid alarms and it became a mad dash up to the slit trenches. A heavy droning noise heralded the approach of dozens of twin-engine bombers. They unloaded their cargoes of bombs, which shrieked down like arrows to smack the earth, buildings and palms in successive blasts of fire and smoke.

After carrying this out, the bombers descended in lines like big silver seagulls and the gunners, front and rear, raked Rabaul with heavy machine gun fire. One of the bombers was hit, swerved violently, then went into a shallow dive directly into the jungle-clad hills. Above the ear-shattering noise, loud cheers rose from the men who were taking such a battering.

Just at that moment, Tal felt a terrible blow strike his back, which catapulted him forward to land face first into a group of soldiers. "You okay mate?" came the anxious query.

"Yeah, I feel okay, but my ears are ringing like hell," replied Tal as he scrambled up to regain his position in the trench. What he saw was a great smoking hole about twenty feet away and about the same distance across. "Where are the other blokes that were there?" he shouted. One of the men silently pointed to an object draped on a jungle vine of pretty yellow flowers. The flowers were splattered with crimson, and the shapeless object proved to be all that was left of a man, smashed beyond recognition.

Tal flopped down, suffering severe shock. He was too stunned to be fully aware of the events of the next five hours of prolonged, severe attack.

The large ship that had been loading copra further down the bay was hit repeatedly and became a fiery smoking torch as it drifted helplessly. Tal said a short prayer for the poor souls who were dying about him.

The burning ship came in close to shore and all night lit up the rain-sodden clouds and the wreckage of a once beautiful paradise.

Early next morning after a breakfast of bully beef, Tal was collared by the sergeant major, "Come with me lad, we've got some business to attend to." He was led to a low building (someone's shed he surmised) and found it to be occupied by a couple of soldiers, Kirky and Robbie, the major, and a serious-looking officer with a red bank on his hat and on his shoulders.

"Tal, this is the brigadier! Sir, this is the chap we were referring to!"

The brigadier angrily smacked the desk in front of him. "This is just a boy. You've got some nerve Major, wasting my time when things are this bad. Get him out of here and send him off to the Mission or anywhere but here."

"Sir," spoke up the sergeant major, "I know the lad is young, but so was I when I joined up in 1915. You know my record since then. I handled it and from all accounts so can this lad. He's willing and able and we need everyone we can lay hands on. He's small for his age, but he's not too small to want to help out. I say give him a go Sir."

"Alright, I leave it to you and the major. Get his details and swear him in, then teach him a bit about the army."

Tal was hustled off accompanied by Kirky and Robbie. "We're your honoured escorts on this most auspicious occasion," said Kirk.

"Where's Dump? Is he okay?" queried Tal.

"I'm afraid Dump is gone, Tal. He caught a direct hit yesterday. We looked for him, but there was nothing we could bury." Kirk spoke gently to ease the shock for the boy. Sadly, the trio followed the officers to a dugout shelter in the foot of the hills.

After giving his name and address, next of kin, etc., Tal was asked, "Date of birth?"

"28th June 1925, Sir."

"You mean 1923, don't you?"

"No Sir, 1925."

"Well, I've got 1923 here and I can't make corrections."

A slight nudge from Kirky made Tal realise the implication. "Sorry Sir, I didn't mean to confuse you." With his face as red as beetroot for being so slow to grasp the significance of the argument, he went on to fulfil his duty.

He solemnly swore his oath of allegiance to King and country and was to puff out his chest and swell with pride as the major put away his pen and said, "Congratulations Private." He was now an Aussie soldier.

Indicating Kirky and Robbie, the sergeant major ordered, "Take this soldier away. Find something else he can wear and then teach him all you know. That won't take very long, will it boys?"

The raiders came as usual that morning, and subjected the 2/22nd to another awesome attack. They had silenced the ack-ack guns and the small arms fire seemed to have no effect. They had destroyed the last of the pitifully outdated airforce, which proved to be totally inadequate against the modern, highly manoeuvrable and very fast Japanese Zero fighters.

All of the soldiers knew by now that all they could do was to wait: wait for the huge army of well-equipped, experienced and fanatical Japanese soldiers. They would be drastically outnumbered and knew they could not survive; but, like those heroic pilots in their outmoded aircraft, they would fight anyway. Surrender or retreat en masse was not considered.

When the rains arrived and brought with it a welcome reprieve, the sad task of finding the dead and injured began. They were taken away for burial, or if alive, to the Aid Post for treatment.

Tal and his two companions, Kirky and Robbie busied themselves deepening trenches and adding camouflage. Kirky began to instruct Tal, who was quick to inform him, "You can't teach me anything about the .303 rifle mate. My dad was captain of the local rifle club for years. He taught us all about it and we learned how to clean and care for it from an expert. It's too big for me anyway." They went on to show him how to aim and fire the two- and three-inch mortar, the complicated Vickers machine gun and the Tommy gun—a .45-calibre submachine gun. Tal admired this one and wished he could have one.

"I'll paddle off to the store Tal and see what I can find for you," said Kirky and disappeared in the rain. He returned about an hour later, carrying a .310 single-shot, which on inspection, proved to be as close to Uncle Jack's .32 as you could get. It was light to carry and according to Robbie, "Good for about fifty yards if you pull the trigger as hard as you can."

In the later afternoon, the sergeant major appeared with a sergeant and corporal in tow. "I want you to pack some rations. Wear full battle gear and proceed across to the north side. Stay in a high position above the beach and if you see any enemy, report back to me on the double." He handed a small item to Kirky, who stuffed it in his pocket.

"You expect them soon, Sir?"

"Yes," replied the S.M. "They'll most certainly be here by tomorrow. A report came in by radio that they're all just outside the harbour. Don't think they'll come in tonight though."

Tal's heart began to race as fear took hold of him and his hands were trembling. The S.M. saw his white face and the symptoms now so obvious.

"Don't be ashamed of fear soldier. I'm just as scared as you. Remember, even a mouse will attack an elephant if he's cornered, so you'll be amazed just how quickly your fear will vanish when you get to have a go at them."

"Thanks Sergeant Major. I'll do the best I can."

"That's all we can all do son. Now keep your chin up. We Aussies are more than a match for anyone on earth."

Consoled in a small way, Tal got together his gear, looking rather ridiculous with two big ammunition pouches in front. A large pack was strapped over his small shoulders. This held his rations and his civvy raincoat and a wide belt around his waist held the precious knife he'd purchased in Brisbane.

With a hat too big for him and a full-sized groundsheet draped over the lot, he caused a lot of laughter from his two mates as he clutched his rifle and was ready to go.

"By cripes Pee Wee, when the nips see you, they'll bolt for sure. Never saw such a dangerous-looking animal in all my life," laughed Robbie.

The trio then began the walk up the wide track that led to the northern beaches.

About a quarter of a mile into the journey, Tal pointed out a path leading off to the left. "My mate Bel lives up that way you know. Wish I was going up there instead."

Finally, long after dark, they reached their destination. The track wound down between two ridges onto an area of flat land covered in coconut palms. They climbed up onto the top of one of these ridges that commanded a view of the bay. Watom Island lay out there a few miles away, but it couldn't be seen for the darkness and rain. "We'll have to sit here till morning, so let's make ourselves comfortable. We can't see anything in the dark and can't hear above this @@!!!! rain," said Robbie.

They managed to shelter themselves by huddling under their groundsheets, and in turn took short naps as the rain poured on their backs and their spirits.

Sometime after midnight, the rain ceased and the mosquitoes moved in. They applied the virulent repellent supplied by the army and sat talking about their home towns, families and the odd remembered tall tale.

Suddenly, a strange sound was heard, "What the hell's that!!" They strained their ears in the silence of the night until Kirky exclaimed, "It's begun. That sound you hear boys, is war." Though very distant, concentration identified the sounds as being from large guns. Lower, yet still audible, the sounds of lesser but constant explosions not unlike muffled thunder reached their ears.

Kirky checked his watch. "It's after two-thirty. There's no movement here, so let's get back there as fast as we can go."

They scrambled back downhill onto the track and set off for their five or six-mile journey at a good jog. After an hour or so, they were compelled by exhaustion to stop and rest. The sounds were nearer now, but a lot less in intensity than previously.

"Reckon our boys may have stoushed them," said Robbie. "Let's hope it's not the other way round," observed Kirk.

The trio was not aware of it, but at the time they had picked up those sounds, the Japanese had begun a concerted attack. The 2/22nd fired flares and as the enemy attempted to negotiate the wire and the mines, they were relentlessly driven away—losing many lives in the process.

As the sky began to show signs of another day, Kirk, Robbie and Tal were still slogging their way through the mud towards Rabaul. Hell broke loose up ahead. They could clearly hear the roar as literally thousands of rifles and machine guns opened up, an odd shell from ships feathered above their heads to explode in the treetops. Mines were exploding and mortar bombs could be heard with their distinctive 'crump'. As they forged ahead, they were able to hear shouts and screams of men in mortal conflict.

"Wait here for a bit, fellas, I'll go and have a look before we barge in," said Kirk and he raced downhill out of view.

"What do you think Robbie? Do you reckon we're winning? Wish Kirky would hurry up," said Tal, his voice shaky with fear and apprehension.

"He's coming back now. Keep your eyes open now. There's not as much racket as there was, so anything could happen."

Kirky joined them again, so exhausted that he was forced to sit and gasp for air. He waved away their questions, as he didn't have the breath to answer. They weren't to know that he was also in shock at what he'd seen.

After recovering a little, he spoke, "It looks like we've lost fellas. There are dead men as far as you can see, and there are Japs everywhere. The only thing we can do is surrender. They said in their leaflets that prisoners would not be harmed." With that, he stood up, grabbed Tal's rifle and threw it away into the thickets. He opened Tal's ammo pouches and threw away all of the bullets. He then took from his pocket a strip of white cloth with a red cross and tied it firmly on Tal's arm. "May get you out of trouble little mate. Now we'll have to go down and give ourselves up or we're goners for sure. Remember Tal, you are a non-combatant. You are part of a medical team and got lost. We found you."

The trio came down out of the bush above the slit trenches. Tal was to look upon a sight so horrible that it would burn its way into his memory for the rest of his life. Japanese and Australians were lying everywhere he looked, most of them sitting wounded and crying in agony, others twitching and shivering in death throes. Japanese were moving about the trenches, firing shots or driving their long bayonets into helpless wounded Australians. It was obvious that all resistance was at an end.

From a trifle behind Tal and Robbie, Kirk said quietly, "Don't move a muscle, we have some nasty company."

About six Japanese soldiers jumped up suddenly from behind the wreckage of a fallen tree and with great purpose pointed their rifles.

"Don't move Robbie," was all Kirk could say. One of the soldiers fired his rifle and Tal felt the passage of the bullet pass his face. He thought, 'If you did that to frighten us, then it worked.'

He heard a thump and a sound behind him and glanced round to smile at Kirk. To his absolute horror, saw his good friend thrashing on the ground as his smashed lungs fought for air and life. In a moment, Kirk was lying still in death.

"You bloody mongrel!" roared Tal. "He was a damn good bloke and he wasn't about to harm you! You bloody mongrel!" The only effect this had on the Japanese was to cause them a deal of satisfaction. They laughed in his face and mocked his outburst.

One in the group had some insignia on his cap and shoulders. He stepped around the fallen tree and confronted Tal and Robbie. To their surprise, he spoke excellent English. To Robbie, "You will very carefully lay down your rifle." This was done and he went on, "You will bow to honourable Japanese soldiers like so." The two did as was ordered, neither being in any frame of mind to resist.

"Your comrades fought with much courage, but as you can see, they were very foolish not to have heeded our offers. You are now our prisoners, and will follow every order which we give you."

To Robbie, "You have been given orders to regroup in the event of defeat. You will tell me the place of your rendezvous, now."

Robbie looked him straight in the eyes and said, "Get stuffed. Expect me to betray my mates? Would you betray your mates? The rules of war forbid me to assist you in any way, so I repeat, get stuffed."

The officer gave Robbie a slight bow, and Robbie returned same with a smirk.

After a brief burst of orders to his men, they came round and herded Robbie away to a distance of fifty yards where they stopped and waited. Tal could just see them through the leaves of the downed tree and hoped he would come to no harm.

The officer turned to Tal, offered him a cigarette, which was declined with, "Thanks, I don't smoke Sir."

"You will answer for me the question I asked of your friend. He is foolish as we do not intend to harm you."

Tal, having so little knowledge of the rules of war, the Geneva Convention, and not with the benefit of hindsight, offered the only piece of information he was aware of. "I think we were supposed to get down south of Vulcan, but we were told it would be every man for himself."

Seemingly satisfied, the officer left Tal in the charge of just one soldier. He moved off toward the other soldiers with Robbie, removing his long shining sword from its scabbard as he went. The whole group disappeared from view up into the mouth of a gully and Tal was never to see Robbie again.

Most of the Japanese troops had moved further down what was left of the town and Tal had for company just one soldier. On studying him, he decided there wouldn't be much age difference between them and there was certainly no difference in height. The rifle the soldier was carrying was as tall as he was and with the addition of a two-foot-long bayonet, looked ridiculous. Tal thought to himself, 'I'm not frightened of this character. I'll bet his mama doesn't know he's missing either.'

The young soldier had also been studying Tal, and as if he'd read his mind, heaved up the great rifle and made motions with the bayonet while uttering short guttural gibberish. Tal could not, of course, understand the meaning of this behaviour, and held out his arms and shrugged.

His tormentor began to become angry and began short jabbing actions with the bayonet as he shouted his unintelligible orders.

Tal hopped sideways over a fallen coconut log and the soldier followed him. Tal stopped with his hands out in front of him. The young soldier, crying out in a fury now, made a lunge. Tal felt the bayonet pierce the skin of his stomach and completely lost his temper. He dropped his hands onto the rifle and tried to wrestle it away from this idiot who was no doubt trying to kill him. He moved past the bayonet and gained a firm grip of the rifle. He could see, face to face, the alarm of his opponent, but decided very firmly he was not about to be killed. Still maintaining his desperate grip, he suddenly pushed. The young Japanese crashed backwards over the coconut log and lost his hold on his rifle.

Tal then experienced the phenomenon of how fast the brain collects and stores information that would take ages to tell. Every movement was in slow motion. He saw the young man's panic as he lost his hold, his face screwed up in fear, as he knew that this was the end for him. Tal seemingly slowly swivelled the long rifle around until he held it firmly in his hands. Just as slowly, he seemed to clear the ground and as he came down, he saw the bayonet pierce the uniform, felt it scrape past a rib, slide through the chest cavity and past another rib down into the earth. He jumped up on the long rifle and drove the full length of the bayonet as far as it would go. The young man did not utter a sound, but his face, full of fury moments before, changed into an expression of sadness as he looked up at Tal. It was as though he offered a final forgiveness. Feelings of horror and pity surged through Tal's mind and he tore his eyes away from a face that would haunt his memory for as long as he would live.

Stirring himself back to reality, expecting the whole Japanese Army to fall upon him, he saw with relief that no-one had seen what had gone on behind the tree. He moved like a lizard back behind bushes, through the thickets, past the trenches of death and into the bush by the path that led back the way he'd come with his mates. He said to himself, "This also leads back to Bel." He looked back at the scene and saw the long rifle still standing upright over a small bundle that could well have been himself.

He felt a stinging in his stomach as he hurried alongside the path that wound up toward the village. On looking down, he saw that his clothing was red with blood from his waist to his feet.

He noted with relief that there appeared to be no signs that the Japanese had preceded him, so he took off his boots, threw them into the jungle and managed to run.

Bel's mother was the first to see him as he staggered with exhaustion toward her home. She placed her hands over her face and screamed, then ran to Tal crying, hugging and patting him as she helped him inside. "Poor Tal, poor boy. You no die finish. Me go get help. You stay still. Poor boy," she cried with tears literally streaming down her face. "Don't worry, Mother," said Tal. "Not as bad as it looks. I'll be okay. Can you bring Bel and Joseph?"

Bel's mother was away for quite a long time. Tal laid himself down on the mat he'd used before and thought deeply about all he'd been through. He'd been prepared to take part in battle and to kill Japanese uniforms, but here he'd taken the life of a young man like himself. He decided it would be the most disturbing personal contact he would ever experience. He asked himself was this the courage he was seeking, could this be called bravery or was it an act born of anger. He was trying hard to reason within himself as he drifted off into exhausted sleep.

He dreamed a distracting mixture of his lost mates, his home, and of the horrors he'd seen.

A small sound woke him and a fearful cry escaped him. For a while, he couldn't familiarise himself with his surroundings or identify the kindly worried faces that gazed upon him in his pain and fear.

"My good friend, Tal. We have you back. Please do not attempt to rise," came Joseph's voice in the semidarkness. "We shall attend to your wounds now that you are awake." With that, Bel and his father carefully stripped Tal of his clothing and washed away the blood and mud from his tired body. Joseph studied the wound and announced, "You are extremely lucky my boy. Your skin and the inner wall have been pierced and I can see a small area of intestine. It is very red but has not been broken. We must make haste to cleanse it of infection."

They draped his lower body with a white cloth and Bel's mother came in with a bowl and strips of cloth. She liberally doused the wound with iodine and then rubbed salt vigorously into the wound. "Salt help stop blood quick time Tal," she consoled him as he squirmed and groaned with pain.

When she was finished, Tal began to sob uncontrollably as she placed more salt on the wound and bound his stomach with clean white cloth.

"Do you feel the pain so badly?" asked Bel over Tal's sobbing.

"No mate," he managed to reply, "I'm overcome, so overcome." His sobs preventing any more.

"It is not good that you lie still for too long Tal. You will soon become too stiff to move. You must get up now and move around as much as possible. You are in very much danger here," advised Joseph, as he and Bel helped Tal to his feet, adjusted the lap-lap and gently walked him around and around the main room.

After a while, he regained his composure and red-eyed with tears, thanked his three friends and saviours. He shook their hands and put his arm about Bel's mother's shoulders, which served to bring on another bout of crying from them both.

"While you are here, you are one of us. After you have left, you shall still be one of us. We all love you Tal," was Bel's emotional contribution, which was quickly affirmed by Joseph. "Enough of the tears everybody. We have much to talk about and there are plans to be made. We shall have an early meal and then find out what has to be done."

During the meal, of which Tal ate very sparingly, he learned that Bel and his father had been out on a high hill overlooking the town and had seen everything of that terrible morning.

They were aware that the Australians had been badly beaten, but that maybe a hundred or so had moved away towards the south, a lot in lorries and others on foot after them. He was not aware of the fact that the 2/22nd Battalion of some 1600 men had killed over 5000 Japanese and had wounded many hundreds more. This was not to become known for another three and a half years. The 1600 Aussies had become less than 150 in the space of a couple of hours.

They withdrew from their lookout about midday and were well on their way home when they were met by a very exhausted and tearful mother.

After dark, they sat on the verandah and discussed the day's events. Tal told them of the dreadful experience he'd endured and the awful guilt that lay heavily upon him. "I could have knocked him senseless or tied him up, but I became an animal and killed him. Do you think I've done wrong Joseph? What would you have done Bel?"

"I speak for both of us Tal," said Bel. They killed your mate for no apparent reason and I'm sure that soldier was left behind to do the same to you. It is not wrong to kill in self-defence. I would have done exactly as you have done," said Joseph. "Had you been in the trenches today, you would no doubt have killed many more than one man, but you would also have died. Give thanks to God that he has cared for you this day. You still have your life ahead of you Tal, and we must endeavour to see that you are spared for the future."

"I've got to try and join up with the others," said Tal. "If I stay here, they'll find me and then you will be in bad trouble. You've all been so kind to me and I'll never forget you. I'll get going in the morning. You said I should keep moving and that's what I'll do."

"You'll have to be extremely slow and careful my boy, as we also saw groups of Japanese moving down the south road. They may catch up with your comrades and they may not, but you must be on constant alert at all times."

After Tal lay down to sleep, he heard the little family talking in low tones in their own language. Their voices rose and fell and he could feel that an argument on a grand scale was in progress. They went on far into the night and were still talking as Tal drifted off to sleep.

THE LONG MARCH BEGINS

He woke before dawn, shaken awake by an earth tremor, walked softly out on to the verandah to find rain pouring down so densely that he guessed he wouldn't be able to tell north from south. As it brightened into morning, he was unable to see more than ten feet.

"It's going to be some day for a walk Tal." The voice startled him and he found Bel at his side. "We are going to have an early breakfast so would you please join us?"

Tal followed him to find Joseph in his best outfit and Bel's mother beautifully adorned in her lap-lap and 'mother hubbard'. "That's the material we bought in Brisbane. You look really great," praised Tal. "You both look very handsome."

Over a breakfast of hot pancakes and golden syrup, followed by tea, Tal asked, "Why haven't you dressed up Bel?"

"Because, my dear friend, I have decided after much deliberation between myself and my parents, that I shall go with you. You are going to need a trustworthy guide and I am your man."

"Bel, I couldn't ask you to risk your life for me. I'll be okay."

"But Tal, there lies the difference. You did not ask me. I have made this decision entirely of my own accord. If you don't want me with you, I shall follow you at a distance of a couple of yards. So, you see, you have no choice. I am quite determined."

Tal was forced to accept. Bel took up a small sack of food, hung a long sharp machete on his belt, and proudly ordered, "It's time to leave. Quick march!"

Joseph intervened, "We must join hands in prayer. Dear God," he prayed, "take these boys, these men, whom we love, into your loving care and preserve them until we shall come together once more."

They all clasped each other in farewell and the two 'men' left the home and headed out into the cool pouring rain.

Overcome with sadness, Tal followed his equally sad soul mate in silence as they plodded away up the sodden track that led away from Rabaul and the nightmares that would haunt his dreams for the rest of his life. He little knew that he would face other horrors in the days and weeks ahead.

Although his stomach caused him a good deal of pain initially and forced him to walk slowly, the exercise gradually reduced his discomfort and Tal was pleased to be able to allow his great protector to cease supporting him.

The uphill journey was still very slow owing to Tal's condition, but by the end of the day, they managed to pass the last of the other four villages, skirting each one, for as Bel explained, "It will be best for you not to be seen, as the Japanese will no doubt visit here and ask questions. Our people cannot tell of what they don't know. If you would rest here, I shall visit my friends and procure some more food as we have a long way to go. I shall return in an hour or so."

Tal huddled himself down beneath a large tree, the roots of which proved to be of good cover. The constant falling of large drops of water from above proved to be annoying, so he decided to keep himself useful. He gathered up a bundle of sticks, crisscrossed them from one large root to the next, laid a large amount of leaf litter on top and was pleased to see that it would be large enough, and would keep both of them dry for the night.

Darkness began to close in and Tal felt very alone and mournful as he went over in his mind the terrible events that had taken place, but confident that Bel would see him safely through. He was beginning to worry about Bel when he suddenly and silently appeared out of the darkness and rain.

"Well, little friend, you've done an excellent job. You were hard to find in such a good hide. Here, I have sufficient food for at least three days, so let's do justice to it and have a good nights' sleep. I have a tin of matches and a small candle too, so after our meal, I shall have a look at your wound."

The meal of chicken and tough type of scone proved to be fit for a king for the two hungry fugitives. They decided to ration the remainder, for as Bel explained, "We shall be in deep jungle and food is non-existent until we reach the coast."

Tal's wound had wept only a little watery blood and Bel proceeded to douse it liberally with the iodine provided by his mother. "It doesn't sting so much now, mate, so it must be getting better already. Still fairly sore though."

The two settled down on a thick carpet of wet leaves, blew out the candle and soon slept the deep sleep of the physically and emotionally exhausted.

Tal dreamt of the terrible events prior to his escape and awoke screaming obscenities as he re-lived the death of his friend, Kirky, and the fury of having killed the young Japanese soldier. The soft voice of Bel brought him back to reality and he sat shivering and sweating as his trembling body relaxed.

"You have experienced too much for one so young as ourselves, Tal. I have heard such language among white men, but not from you before now. I do hope you don't corrupt me with such a display of vulgarity," smiled Bel as he did his best to bring Tal back to reality.

They peered out from their cover and found visibility down to almost zero. A thick mist cloaked the dripping darkness of the forest. There was no birdsong in the high canopy and the oppressive silence seemed ominous compared to the more open atmosphere alongside the tracks up to this place.

It was quite cold and Tal shivered in his damp clothing. He began scratching at his arms and legs and upon inspecting himself, found them wet with blood. A dozen or more triangular cuts showed where leeches had gorged themselves while he slept. He bent to remove one from his leg and was struck with pain in his wounded stomach. He slowly and painfully brought himself upright. "Bel, I'm really hurting. I don't know how long I can go like this. I think I'm in big trouble." Tears of pain showed Bel that Tal was in deep distress.

"It could be that you have stiffened up during the night. Maybe if you were to exercise yourself slowly, the pain would lessen," said Bel. Tal proceeded to bend slowly forward and back, side to side until he could feel that his wise friend was right. The pain, though making itself very much felt, became at least bearable and he was soon to announce, "As soon as this fog clears we should be on our way Bel."

"We must not wait that long, Tal. All we must do is travel upwards along this ridge. We must travel well up into the Baining Mountains before we head south. If we go down to the lower country, then we shall surely meet Japanese. There are hordes of them and they'll get us most surely."

Tal saw the wisdom of this decision.

"You're the boss, Bel. This is your country and I know nothing of it. I can find my way home like a hungry dog back home, but this is totally beyond me. I wish you didn't have to put yourself in danger for me though. I feel very guilty about you leaving your mum and dad just to help me."

"No negative thinking, Tal. I always do what I know is right and I know I'm not wrong in this. Now while you exercise, I shall show you my prize."

Reaching around the shelter, he produced a longbow and half dozen or so very dangerous looking long arrows.

"These may come in handy in order to procure food. No intelligent man would be seen without this equipment in the jungle. My friends at the village insisted that I take them, though I must admit, I have not had much experience with them."

"Well, I consider myself fairly handy with the bow and arrow," boasted Tal. "We were always making them when we were kids and could often bring down a parrot when we were out camping. They may even help protect us if we meet up with our enemies. Not as good as a rifle, but I'd sure hate to be shot with one of those arrows."

Bel proceeded to dismantle the shelter until there was no trace that anyone had spent the night there and then heeding Tal's condition, slowly led the way upwards through the dark foggy forest.

Tal was surprised to find that the going was easier here compared to the dense growth found in the lower country. The earth had not been bogged up by countless foot traffic but was covered in leaf litter that proved quite pleasant to walk upon. The big tree trunks towered up into the mist and though no daylight could reach them, the two moved silently and invisibly upward in the dark grey morning. Tal found comfort in following this big black friend and was grateful for his never-ending cheerfulness, confidence and caring nature.

Though it was full daylight, the fog did not lift and the two plodded ever upwards, dodging around fallen trunks of trees. Tal could never have envisaged such a terrible place. There was no real light to brighten the forest or their spirits. He was still in some pain and travelled slowly, making the day stretch even longer than usual. Bel was always attentive to his needs, but insisted that to rest too long would impede their progress and only induce more pain.

Toward the end of their second day, when the only indication that it was over was that it became impossible to see in the black fog, they found a couple of crossed tree trunks and made camp. They consumed their small ration of food and settled down on a liberal mattress of leaves.

Tal had never felt such exhaustion and guessed that Bel also was suffering the same, as talk between them was at a bare minimum. It was not long before they fell into a deep sleep.

Tal woke with a start at hearing scufflings in the leaf litter outside their haven. There were strange short breath sounds very close and a chill of fear ran through him as he quietly nudged Bel. Bel listened for only a few seconds, then Tal heard him fumbling in his belongings.

There was a small flash of light and a loud 'woof' from Bel and all sounds ceased.

"Wild pigs, Tal. They've probably never seen a couple of matches go up. They won't return. I can see a little out there, so it must be daytime up top. Now we'll share one of mother's biscuits and be on our way again."

Their drink was obtained by cupping their hands against the trunk of the nearest tree and gathered their fill from the cascade trickling down from the invisible canopy high above.

"My belly feels a lot better this morning mate. Still sore to touch, but I'm not as stiff as I was."

Bel inspected the wound and happily assured Tal, "All the redness has gone, and it seems to be closing up very nicely. I'm sure that you'll be as good as new in no time."

He applied a little more iodine, reversed the dressing and they were on their way again.

The third, fourth and fifth days were a repeat of the others with dense fog and the constant pouring of water from the canopy. It could not be called rain, as it was concentrated pourings that could be dodged. It was, however, very consistent, very annoying and uncomfortable.

Surprisingly, there were no mosquitoes as Bel explained, "They only inhabit more open areas where there is undergrowth and succulents on which the males feed. The female mosquito is the one to fear as it needs blood in order to breed.

We are now high up in the Bainings Tal, and there's no more climbing up for us. We shall work our way downwards but at a slant. We'll cut across ridges, but always on a slant as we approach and then straight down the other side. That way, we shall be moving toward the south-west and Wide Bay."

The jungle of tall black towering trees, fallen logs and the precipitous sides of the terrain made progress extremely difficult. The pair slogged and skidded along the slopes for most of the day until they reached the end of the long ridge, which finally dipped away down into the darkness.

"Well Tal, we have conquered one of our obstacles. Now we shall turn right, go straight down and rest at the bottom," said Bel.

"That's a welcome change, Bel. I'm sure my right leg is at least six inches shorter than my left. What a slope. I bet we didn't manage a horizontal effort as I'm sure I skidded a foot down for every three forward. Anyway, let's have a go at finding bottom while we still have some light. I'm feeling as hungry as a horse."

"Right you are, Tal. Follow me!" yelled Bel, and the two skidded mostly on their rumps, bumping into trees or into each other until it became a noisy happy race for at least an hour. The war was forgotten. They became again, two happy boys enjoying an enormous adventure and it did much to restore their exhausted minds and bodies.

However, upon reaching the bottom, their uplifted morale was considerably dampened as Bel made the sober announcement, "Tal my friend, we must now become very resourceful. We have sensibly rationed our food supplies, but now we have none."

This dilemma completely astounded Tal. "Bel, we've got a long way to go. How are we going to survive in this place? Not a bird or an animal to be had. No villages. What the hell are we going to do?" He sat on the wet soggy ground, not despairing, but feeling a great sense of helplessness. He wondered if his big black mate felt the same.

"Come now Tal," spoke Bel, "remember resourcefulness. We have a small stream here and I'm sure there'll be some creatures in or around it. Let's build our shelter and in the morning, we may find something."

There were sticks aplenty and in no time, they were stacked appropriately, covered with leaves and the two hapless friends huddled together, not by any means dry, but at least protected from the incessant pourings of rainwater that fell from high up in the dark canopy.

Night came down suddenly as always, and a breeze sifted through the blackness chilling both to the very narrows of their bones. Lack of food to warm them caused both to begin an uncontrollable shaking and pure misery became a completely inadequate description of their plight.

"Tal, we are rapidly suffering the effects of this awful cold. We cannot light a fire in these conditions, so we must exercise in order to keep our circulation going," shivered Bel. Rather than move out of their shelter, they together planned an effective method of exercise. Facing each other, they placed the soles of their feet against each other's clasped hands and pumped long and hard alternating between hands and legs. Soon they beat the awful shuddering. The entire night was spent in this fashion. After ten minutes or so of rest, the need to repeat the exercise became imperative.

"This is the coldest night I've ever experienced Bel," gasped Tal. "The weather at home gets to be colder than this, but at least I've always had a warm bed to sleep in."

"It's the altitude of our position Tal. When we are further down toward the coast, we shall not have to endure conditions such as this. I must admit that this is also the coldest I have ever been."

A gradual lightening of their surroundings heralded another day and the two forlorn friends were pleased that at least the seemingly incessant rain had in fact ceased. The world way above in the treetops seemed also to be grateful. The pair sat and listened to a great cacophony of screeching whistling birdsong.

"Love to hear birds Bel, but I'd love to have one for breakfast," said Tal.

"Listen hard Tal!" exclaimed Bel, and they ceased their talk.

"Hear that, again, again. There!" shouted Bel as he jumped to his feet and made a great leap over Tal, causing astonishment and some alarm.

Bel pounced on something on the floor of the forest, listened, and half a dozen times pounced again.

"What in hell are you up to Bel?"

"Come, see for yourself."

Tal hurried to find that Bel had a big handful of small round fruits, being some sort of fig.

"You see Tal, parrots are messy eaters. They waste more than they eat. Watch out and see how many more you can find before they stop feeding."

The two partners finally gathered about a half kilogram and sat down to do justice to this sparse, though welcome gift from the birds.

"You intimated your concern that God had deserted us during one of our talks. What have you to say now Tal?"

"Well mate, it wasn't manna and it wasn't from heaven, but it comes mighty close," said Tal, evoking much happy infectious laughter from Bel. "I'm not complaining," said Tal, "but that just whetted my appetite. Wonder if we can find anything else."

"We could try the stream as I suggested last night," said Bel and jumped to his feet. Followed by Tal, he proceeded to examine the small stream that ran fairly quickly past their campsite.

"I see nothing here Tal, but it could be worth following down for a while." Bel picked up his now empty bag and his bow and arrows and they moved slowly downstream, studying the stream and surroundings very closely.

"No bugs or witchetty grubs Bel! I'm hungry, but not that hungry yet," said Tal.

This stopped Bel and he insisted on a complete rundown on witchetty grubs for he had never heard of such things before.

"I shall remember nevertheless Tal. One never knows how things may turn out for us."

A few hundred yards of searching, found them clambering under giant fallen trees where the stream spilled over rocks into a pool of no great size, but showing some promise. Bel excitedly pointed out four or five small dark fish measuring about three inches in length darting about near the edges of the pool.

"Now, there is food, if we are able to catch them. Then again they would not be here if there was no food for them, so it is logical to assume that there is also food for us," said Bel, as though talking to himself.

"Righto Sherlock Holmes, let's see if you are able to deduce ways and means of satisfying your logical assumption," quoted Tal as he emulated Bel's, as always near perfect, use of English.

"You mock me Tal. I am after all doing my utmost for us both, and feel that such a remark was uncharitable."

"Don't get off your bike mate. I was just having a joke; and I can't help it if my face doesn't do justice to my sense of humour," said a peeved Tal.

At that, Bel broke into loud laughter that rang in the stillness of the jungle. "Quiet mate, there could be Japs around," only brought on more laughter and the unsettled feelings melted away like the bubbles in the stream.

"Aha," said Bel. "Look at this Tal. We shall have a meal of these little fellows." He held up a three-inch shrimp, took off the head, peeled it, popped it in his mouth, swallowed and beamed his big white smile.

"Fair go Bel, not raw, good Lord, we can surely wait 'till we can cook them."

"Not necessary Tal. We did not cook the fruit earlier and there is no need for us to cook these. Just pop them in and swallow. They don't need to be chewed in order to do you good. We can catch enough of these to really fill us up. Like you would say, 'have a go'."

Tal didn't feel well about the first one or two, but eventually settled down to catching them with as much dexterity as his friend. After catching all they could from the pool, they moved off to another and spent most of the morning flicking them out with their hands and eating them until both sat back completely satisfied.

"I vote we have a day off Bel. I doubt if I could walk too far after such a party," said Tal, and was pleased that his large mate was in complete agreement.

They returned to their camp of the previous night, made things a little more comfortable and as the first splatters fell from above, they agreed it had been a good day.

By this time, Tal's wound had healed up well, only bothering him when he was forced to twist or to stretch. This was a result of the outer skin and the stomach wall becoming welded to each other as they healed. This condition was to remain with him for the rest of his life.

The two good friends sat and told each other stories of their experiences over the years of family and friends, likes and dislikes and of all the good things that boys find exciting to see and do. They both agreed that boyhood was now far behind them and that as men, they would do their utmost to honour the teachings and example set by their parents.

They also decided that they would have another meal of shrimp before leaving in the morning.

As darkness fell, the rains had apparently ceased, the noise of falling water became a quiet whisper and the two fell asleep, not to wake until visibility arrived in the morning.

Being earlier this day, their catch of shrimp proved to be exceptionally successful and they decided to carry a goodly amount so that they could be eaten later. "Maybe cooked this time," said Tal.

Another three days were spent traversing ridge after ridge, always a long way to the south and short to the west until on making camp on top of a ridge one afternoon they were startled by a familiar, but frightening sound coming from the next ridge. "That was a blasted rooster crowing," exclaimed Tal. "There has to be someone living there. Should we give them a hoy Bel?"

"No! No!" said Bel with great urgency. "We must be extremely careful. There are supposedly no inhabitants in this area. We could be up against wild kanakas. My father has told me about some of the people he has encountered in his work with the Mission. They completely reject outsiders, even neighbouring tribes and some of them practice cannibalism and head-hunting. We most certainly will not visit that ridge tomorrow, but will be forced to go right down this one until we're well clear."

Tal was more than a little uneasy after hearing this and immediately set about destroying all evidence of their proposed camp. "We'll move off the top of this ridge and set up a good hide on the slope looking their way. That way, if they come for us, we may see them first."

This they did, just as quickly and quietly as possible and settled down huddled under an enormous conglomeration of fallen crisscrossed logs.

About an hour after last light, voices could be faintly heard in the still air. Suddenly it began. A sound that would become familiar—to the extent that it would be hated for a long, long time to come.

"Drum talk," explained Bel. "This is the only means of communication between tribes all over our island. It will go on for about two hours every night. I know a little of it up this end of New Britain, but not further down. It's a little like your Morse code, only pardon me, a little more refined.

The drums are a set of two hollowed logs of special timbers, one longer than the other. When struck in different places they produce various sounds. Various sounds mixed with the rhythm and intervals produce the drum talk. It is a necessary for each boy to read drum talk before his initiation to manhood, and it takes many years of learning to become proficient. I did not have to learn as I was reared at the Mission, so I must accept the fact that in this area, I am considered to be an illiterate."

"What do you think they are on about now Bel? I hope we aren't included in tonight's news," said Tal.

"No, I don't think so," replied Bel, "I can make out a little and it seems as though all the concern is about the war. The comings and goings of men, ships and aeroplanes. 'Baloos' they call them. 'Birds'. They also claim this area as their own and warn others to stay away. Just like the crowing rooster."

Tal listened hard and, at intervals, heard other drums away in the distance. According to Bel, each tribe's news was passed along into the distance with added snippets of gossip and replies were received. This practice was only effective over long distances on clear nights. If it was raining, then the sound would not carry and so all bulletins would be cancelled. Despite Bel's assurances, Tal spent the night wide awake.

Being in a strange land and in the proximity of possible mortal danger, Tal listened to every sound and often woke Bel when some furtive move caused him alarm. Bel would listen and comment, "Just a wild pig, a bandicoot or opossum," then he would go back to his sleep.

Morning light brought relief to Tal's fears. He was happy that at least in such a fearsome situation, he could see. Bel woke with his usual big smile and together they listened as the village came to life. Smoke from their fires drifted across and soon the smell of food made them realise that once again they were hungry. "I'd like to get hold of that rooster Bel."

"I'd like to remove ourselves from this place instead," said Bel as he gathered up his bag and weapons. "You carry the machete Tal and I shall lead. If we meet anyone just stand still and I'll do the talking."

The two men set off at a good pace, hastened by the possible danger of being discovered as trespassers. "They gave out their warnings last night," said Bel, "so we don't need any trouble."

They hadn't gone more than one hundred yards when both saw an awesome apparition at the same time. They stopped dead in their tracks and stared. A completely naked man and his woman stood also very still and stared back at them. The man carried a small woven bag on his arm, the other holding a long spear. His woman carried a large, long basket on her head. "Now they know we're here Bel," whispered Tal. Bel said nothing, but held up his hand as a sign of peace and placed the end of his bow on the ground. The other man did the same with his spear and Tal let out his breath in a long sigh of relief."

"Stay back Tal. I'll see if they'll talk to me," and Bel slowly walked toward the pair. The woman looked very scared and was about to run, but her man spoke to her. She sat on the ground and peered from behind her basket. The two men exchanged a few words, then squatted down and proceeded with a long conversation making much use of their hands as with their speech. Tal listened to their rapid talk, but could not make out anything of what was said.

Tal received an awesome scare at the next turn of events. Suddenly, the jungle rang with loud shouts, whoops and cooees and a dozen or more naked and very threatening looking savages had them surrounded. The air was filled with their rapid chatter and they looked prepared to use their spears and arrows at any moment. The man who had been talking to Bel stood up and called very loudly over and over to the excited throng until they finally quietened down and listened to what he had to say. Bel remained in his squat position and Tal stood rooted to his spot while the man shouted long and loud. His talk was interrupted many times by individuals in the crowd, but finally after about fifteen minutes, he gained their silence and they all listened while he shouted and gesticulated, pointing at the two unfortunates, who had not invaded their territory, but were lost.

Tal shakily looked at Bel, who gave him an encouraging smile and he felt that Bel had achieved safety for them. This assumption proved correct when the savage men put down their weapons, crowded around and listened as Bel told them the story. The whole event took over an hour, but to Tal, it seemed an eternity, as he had no way of knowing what was being said.

Finally, one of the men shouted loud and long and, in a few minutes, there appeared many women carrying large baskets on their heads, young boys and girls, even the odd small child. All were quite naked and Tal was to become extremely embarrassed. He didn't know where to look and he felt like running away. Finally, Bel came to him, "We are indeed fortunate Tal. These people were heading for their gardens. The man to whom I spoke interceded on our behalf, or our travels could well have been over. Some of the men are against us, but we have been given permission to accompany them to their gardens, obtain some food and leave as soon as possible. I have given my word that we shall not speak of this to any other natives we should meet."

It was a relief to feel the fear ebb away and they joined the noisy throng as they moved down through the jungle and up the next ridge.

Here was a clearing of about five acres, fenced with sticks placed closely together. The gardens were full and well-tended. Vines of fruit were growing on trellises and a few small coconut palms were bearing large nuts. A nice grove of pawpaw trees was loaded with long green fruit and a fairly large area was covered in sweet potato vines and peanuts.

A loud hullabaloo commenced at the far end of the garden. On investigating, Tal found that two wild porkers had been caught in a most ingenious fashion. A small stick yard had been erected with an opening for the pigs to gain entry. A few scraps of vegetable were used to entice them to squeeze past two vertical rows of sharpened bamboo pointing inwards. Once in, they were unable to escape those sharp spikes and so were caught.

The men and the children did not work, leaving all the digging and carrying to the women, a practice Tal thought was more than a little unfair. The children were a delight to him. They had never seen a white man before and would stand and stare, with large eyes missing nothing. Tal sat on a log, smiling at their wonderment and soon won their confidence. They crowded around, feeling his fine, fair hair and rubbing his hands and arms to see if the colour might come off to reveal black beneath. They were finally chased off by one of the men, much to Tal's disappointment.

Tal and Bel were invited to take their fill of pawpaw, five corners—a yellowish fruit tasting a little like plums—coconut milk and raw peanuts. The coconuts were broken open and the white, sweet flesh was chewed up with much relish by the two hungry travellers. Bel's bag was filled with peanuts at the lower half, coconut and finally five corners. He offered profuse thanks to the 'luluwai', the man originally met, who, fortunately, turned out to be the chief of the tribe. He accepted the gratitude and then in no uncertain fashion pointed the way out and intimated that departure was the order of the day. The two friends were not about to argue and hurried away, watched by every eye in the crowd.

Down the long ridge they went until they were some hours away from this awesome adventure.

Stopping for a well-needed rest, Bel decided to take a look at the countryside by climbing an enormous tree covered with strangling roots up its entire length. Tal sat and patiently waited as Bel disappeared away up in the canopy.

"Tal!" he yelled from above. "Do you think you can climb up here? I have something to show you."

"I can climb! I'll be right up!" and Tal clambered easily up the long network of roots until he'd reached Bel. Standing up on a liberal platform of branches, Bel pointed off to the south. Tal scrambled to his side, stood up and was struck dumb with the sight. The sky was clear except for the far horizon, and away down over the treetops and ridges was the ocean, blue and beautiful. They stood for a long time marvelling at the birdlife, the blossoms and the beauty of it all compared with the semidarkness down below them. They also planned to follow this ridge down to the sea, still a good two days away.

At long last after enjoying the couple of days of no rain and comparatively easy going, they stood in a fairly open area overlooking the Pacific Ocean. It was indeed peaceful, with sea birds circling above and below, a cool breeze blowing in from the water and the reefs of coral showing up like flowers along the shoreline.

"How very beautiful Bel. It could be paradise except for what could happen down there."

Bel agreed that it was indeed a beautiful sight, but also commented on the dangers that would no doubt await them. "We must be extremely quiet and vigilant from now on Tal. I believe I know where we are as I have sailed along this coast with my father and Father John on the Mission boat. As the land lies, I believe we are at the eastern end of Wide Bay and that Tol Plantation lies away along the shore. It will take us some time to reach, but we shall now be able to procure food. There are quite a few villages from here on and most of the people have had contact with civilisation."

The two proceeded to wend their way downwards until they reached a well-worn, wide track that had obviously seen much traffic very recently.

Bel studied the ground closely and announced, "Lots of boot tracks and others which puzzle me. They are light and appear to be made by a softer shoe. See, the toes appear to be split. My opinion is that they've been made by Japanese. We shall have to keep an extremely good lookout, in front and behind us from now on."

As they moved warily along, staying close to the undergrowth, it became very obvious that the Japanese were indeed ahead. A worried Tal followed his big mate along the track that snaked around the ends of steep ridges, down and across small streams and up sticky, muddy climbs. At the top of each climb, they would sit, watch and listen for any sound that might signal the presence of danger. This proved difficult during the day as the noise made by birds and the sounds of the sea drowned out almost everything else.

Daylight once again began to fade and they decided to look for a safe place to camp. Moving off the track they decided not to go down near the sea, but up a water-worn gully and under a large tangle of vines and other undergrowth.

There the mosquitos moved in on them and they began to suffer.

"Rub mud on your face, hands and legs Tal," was Bel's concerned advice and this proved to be most effective, yet these annoying pests always managed to find somewhere to sting. An hour or two after dark, the mosquitos left and they were finally able to spend a reasonably restful, but hungry night. Their food had been hard to find on that long ridge and to search the small streams they had crossed would have been much too dangerous.

The drums began again soon after the mosquitos had left them and Bel was able to glean a little information. "We are a small item on the news tonight Tal. Let's hope that no-one translates to the Japanese. They mention soldiers again so we know they are ahead of us."

They huddled together half-asleep and half-awake until morning. As they furtively sneaked out of their cover at dawn, they were alarmed to hear voices coming from the far side of a crest and from the direction in which they were to go.

Bel hastily signalled Tal to withdraw to their hide. He remained on the track, looking big and brave, weapons in his hands, still covered in mud, to await whoever it might be.

All too soon and very close, Tal heard the voices suddenly cease, then a challenging call answered loudly by Bel. Tal didn't move from his position but listened hard. The voices did not sound to be as aggressive as experienced earlier and after about thirty minutes of animated chatter the newcomers moved off, heading up the coast towards Rabaul. Bel called to Tal and he was delighted as Bel proudly showed him the contents of his bag.

"They were all workers from a small plantation along this coast. It appears that the owners had fled by boat and left everything behind. The workers were all heading back to their villages and before leaving, were able to salvage plenty of food and other belongings. They were able to hide from the Japanese, but say there are a lot of them. A large group of Australians had passed the plantation earlier and the owners had taken some wounded soldiers with them. They are well ahead of us, but the Japanese are between us."

Bel's bag was full of tins of baked beans, corned beef, condensed milk, rice, split peas and a small loaf of damper-like bread.

They decided to eat only when forced to and to make these rations last as long as possible.

They did however, open a tin of beans and ate the small loaf, as their hunger proved to be too much to resist.

"We must not dally in a place like this. It could prove to be extremely dangerous," said Bel. "We must travel along this so-called 'road', but not on it. We shall stay a short distance from it and follow it down. It is only about ten miles to the plantation so we shall be there by tomorrow, though I dread to think what we may find there."

The two plodded through and around dense undergrowth for many hours until they were able to find easier going back in the true rainforest where it proved to be easier. It was not necessary to cut a path for themselves, but traversing up and down the ridges proved to be exhausting. By nightfall, in the pouring torrents of water from the canopy, they assembled their 'camp'. They relished their new-found rations and slept soundly, a first for both of them for what seemed an eternity.

In the warm wet jungle, the fireflies and the masses of phosphorescent mosses on the ground held no fascination to keep them awake as they had done earlier.

Tal dreamed of home and family until the face of the young man he had killed intruded his mind. He woke with a start and found himself breathless and trembling. Sitting in the darkness of the shelter, he listened to the mysterious sounds of the jungle and the comforting sleep sounds of his big mate and protector.

An hour or so later, light began to outline the big black trunks of the trees. Bel came awake and they shared a tin of baked beans for breakfast. "Food for the soul," was Bel's boisterous thanks for the meal.

"For the soles of your big, flat feet," replied Tal, bringing back laughter into their troubled and dangerous lives.

After four or five hours of plodding up and over the ridges, they came upon a stream that had cut its way deeply through the jungle. It was flowing fast and too dangerous for a crossing. Added to the swift tumbling torrent, the precipitous and slippery banks precluded any thoughts of making a crossing.

"Well mate, I don't fancy going way up again to find a crossing," said Tal. Bel agreed and they moved off downstream, aware of the dangers involved.

Within an hour they had left behind the protective gloom of the deep forest and were pushing their way through a massive tangle of ferns, palms and liana vines, all of which tore at their skin. Leeches were hanging on every second leaf and proved to be as irritating as the mosquitos.

"We are nearing the road again," whispered Bel. "We must be

extremely vigilant and very quiet. We must also be very close to the Tol Plantation. The Japanese will most certainly be there. I have a plan that may work. We shall work our way closer until we reach there and I shall put my plan to the test."

Tal was curious, but on Bel's insistence of being quiet, he decided to wait and see for himself.

The undergrowth thinned out, showing evidence that it had been cleared not too long ago. Regrowth is fast in the islands. Still, it was easier to penetrate and still afforded good cover.

The stream had now almost reached its destination in the sea. The waters had spread out and become shallower so the pair made their crossing.

Skirting the area just described, they detected a faint but obnoxious odour hanging in the still air. As they proceeded, the odour became the foulest stench imaginable. They came upon the source together and fell back in horror. The decomposing body of a now unidentifiable soldier lay bloated and festering in the heat of the day. He had been wounded, as bandages were visible in the mire of his throat and on his hands, which were uplifted in death, giving the appearance that he was reaching for supplication.

"I'm afraid we shall have no time to bury this poor soul Tal. Let's just say a prayer for him on our way," said Bel. They hurried back into more dense cover and as they paused for breath, both began to retch uncontrollably.

As they began to move off, Bel stooped and held out a .303 rifle and a small but reasonably heavy haversack. "Looks like whoever owned them, has had to abandon them," said Tal.

Bel added, "A rifle is of no use without ammunition Tal, so we shall also abandon it. What do we have here?" He opened the haversack and saw what proved to be twelve grenades.

Tal grabbed a couple, screwed off the butts and exclaimed, "We won't abandon these Bel, see, they're primed with four-second fuses. These may save us in an emergency."

"But Tal, they're so heavy. They'll slow us down if we have to move quickly," said Bel.

"I'll carry them myself. If we have to run in a hurry then we can throw them at whoever is after us. I was taught about these things and I'm not leaving them behind," said Tal.

They then proceeded very cautiously until they finally saw the undergrowth give way to a flourishing grove of coconut palms. "This is Tol Plantation," said Bel. "About a half-mile away and down near the beach is the owner's home and that of his workers. We shall have to find a good place to hide until we find out what we can do next."

TOL PLANTATION

The two walked back to where the flat ground gave way to jungle and near to the stream they had crossed, making sure they stayed well clear of their grisly find.

They located a very large tree on the steep bank. It stood at a precarious angle and Tal guessed that after not too many floods, it would finally come down. They parted the scrub around its base and found a cavity amongst the roots suitable to their needs. When the scrub was pushed back it was completely invisible.

It was by now close to dark, and the rain, so punctual every day, was pouring down and into their hide, making things extremely uncomfortable. The stream rose to a raging torrent and they hoped it wouldn't rise high enough to further erode the banks and topple this great tree. They had a meal and as there was not sufficient room or a dry spot, they resorted to taking turns to sleep sitting up.

As dawn was breaking, Bel unfolded his plan. "I must abandon my modesty and become a 'kanaka' today. I'm going down to the beach, see the Japanese and find out what I can about your friends, and see if any possibility exists that will see us out of our troubles. I would like you to stay here Tal so that I shall know where you are. If we do otherwise, we may become separated permanently."

With that, Bel proceeded to turn himself into a wild jungle man. He removed his lap-lap and whittled a couple of sticks, which he pushed into his curly mop of hair. Another he carefully whittled with the aid of Tal's pocket knife, sharp at each end, a notch in the centre, cut almost through. He bent each end making each crack open and then spring back. This he clipped under his nose and he gathered up his longbow and equally long arrows.

"Wish I had a camera Bel. You're a really frightening sight," marvelled Tal.

"I may have to be away all day my friend. Wish me well and hope that I return with some good news."

With that, Bel strode off, not carefully or trying to hide himself, but to all appearances a wild warrior investigating the activities of strangers in his country.

Tal settled back and prepared for the first long day without his loyal friend. At times, he ventured to put his head out and listen to the sounds of the jungle and to try and identify those sounds.

At irregular intervals, he heard rifle shots in the distance, muffled by the jungle, but quite distinct and he prayed that his mate was safe. The only food he allowed himself for the day was to chew on dried coconut that had been given by the plantation workers the day before.

The day was warm and sunny, the stream had settled down, clear and inviting and he wished he could get down to it and wash away the mud from himself and his sparse clothing. His hair was matted and becoming long and he was due for his first shave. "Crikey, I think I'm getting old," he told himself and this little chuckle, he felt, was long overdue.

The anxious day passed slowly for Tal and in the late afternoon, his long watch and wait was rewarded. Bel, still the wild kanaka, appeared carrying a bag and still retaining his weapons.

"How'd you go mate? Did you see our blokes? Are they OK? How many Japs?"

"Steady on Tal, all in good time. We're in danger here and have to move on soon. What I have to tell you is not good. I shall tell you everything and we shall make our decisions together." Over a small meal, Bel told of the events of his day.

"I left here and headed around the plantation. It was not long before I heard the loud voices of many Japanese. I kept myself hidden from view and was able to hide myself until I came near to the copra sheds. There I saw the most terrible things one could ever imagine. I don't believe I shall ever recover. That men could become so bestial is beyond me. I hate the Japanese even though hate is against all I have ever been taught. I saw Australian soldiers beaten senseless with fists and rifle butts. Poor helpless men, bleeding and suffering, callously stabbed to death with bayonets and swords. They tied some to palm trees, made them watch everything and then the Japanese took turns stabbing them in non-vital parts of their bodies until they were almost dead. Then they would shoot them through the chest and watch them die. They actually laughed together during this awful butchery. It was happening in other places around the area within my sight and I could do nothing. They were making other Australians drag the bodies away into the bush and there the same thing would be done to them. It was the most terrible day in my life and I was petrified with horror and fear. It was not long before I was observed by the Japanese who came running towards me with their bayonets and I thought my time had come. A command from an officer stopped their mad rush to kill and I was forced to proceed to his tent where he tried to question me. He tried Japanese, English and called in another officer who had a good knowledge of Pidgin English. However, I feigned ignorance of these languages and answered all questions with inventions of a language even I could not understand. Through sign language, I managed to make them believe that I had come down from the deep jungles in the north, looking for salt for my village. They pointed to a small group of Australians and wanted to know if I had seen any of them in my travels. Of course, I intimated that I had not and eventually the officer, again through signs, gave me a clear understanding of what would happen if I were to assist their enemies. I was in a state of terror the whole time but managed to stand tall and to give the impression that the sight of death was no stranger to me.

He then let me know that his war was not with our people and that we were free of white dominance. If we respected the Japanese we would be respected and protected by them. He had photographs taken of himself with this savage heathen. Other officers also lined up to do the same. I suffered this in silence and tried to look as fierce as possible. He gave some orders and this bag was brought to me. It contains some salt and about twenty pounds of rice. They then ushered me from the tent and I was free to wander. I saw two barges on the beach and two more arrived each loaded with men and supplies. They were erecting tents everywhere, so I believe they are making this a base camp. As I was moving about, I noticed a big pile of rifles and packs. They were unmistakenly Australian. They are at the open end of the copra shed. The packs have been looted, as I saw letters and photographs lying in the mud—the treasured personal possessions of all of those poor men. Tal, I see no chance of escaping this dilemma to the east or the west. We must do what we ought to have done initially, and that is to reach the north coast. It is only about twenty miles away on the map, but we must travel over very high mountains and into the territory of some very inhospitable people."

"They can't be any worse than the Japanese. We're dead if they find us, so I'm with you Bel, but I would like to have a gun or two. These grenades may come in handy, but I'd still like a gun," said Tal. "I can't throw a grenade as fast and as far as a bullet."

"Right," said Bel. "Tonight, I shall return to the copra shed. The tents are not too close, so I should be able to reach there unnoticed and I shall steal whatever I can."

"Not on your own mate. I'm going with you. After we do this, we're going to make ourselves as scarce as hen's teeth. We won't come back here."

"Agreed Tal. Let's grab our few things and get going. Once we're out of this forest we'll be able to see fairly well."

After moving cautiously for approximately an hour, they emerged on the northern side of the plantation. As Bel had predicted, vision was reasonably good under bright starlight. They came upon a wide track where they placed their meagre possessions behind tall grass and slowly crept toward the Japanese encampment.

Tal was to see for himself the evidence of brutality.

Four coconut palms appeared to have peculiar shapes at the end of the trunk. On investigation, both friends were horrified to find bodies. Their hands had been tied behind them around the coconut palms and they had rope about their throats, also tied around the coconut palms.

They had been stripped naked. Their bodies, from their necks down, were black with dried blood from many wounds, no doubt inflicted by bayonets. Tal recoiled in horror and promised himself, 'One day I'll get even.'

The soldiers were apparently celebrating their vicious victory. Campfires were burning, lanterns were being carried about, loud voices and singing drowned out any slight sound made by the two desperate fugitives.

Tal was amazed that in their guttural language, they were actually singing "Blue skies, smiling at me, nothing but blue skies do I see."

"There's the copra shed Tal," whispered Bel.

As they moved closer, their hopes were dashed. Just inside the shed were two Japanese soldiers inspecting the weapons and packs. In the lantern light, just thirty or forty feet away, their features could be clearly seen and Tal's heart was pounding so hard he felt they must surely hear it.

"We'll wait and see what happens," whispered Bel and they watched as the two enemy soldiers separated rifles and ammunition, Bren guns and submachine guns.

After what seemed an eternity, the task was completed and the two took up their lantern and went off to join their comrades.

"What good fortune. They are so sure of themselves that they haven't even put a guard on here. Let's get in and out as fast as possible," insisted Bel.

"I shall watch around this corner. You select what you think we need and push it out near me."

Tal hurried in; his attention drawn immediately to a small pile of Thompson submachine guns. He had been given instruction on these weapons by his lamented friend, Kirky, and considered them to be much more ideal in these circumstances than the cumbersome .303 rifle.

Conveniently placed beside them, were many rounds and flat ammunition magazines capable of holding fifty .45-calibre rounds. In a box beside them, he spied the waxed cardboard cartons, each holding fifty rounds.

He quickly and quietly took up eight of these cartons, placed them in a large shoulder-born pack, stuffed it into another pack, scooped up the guns and four ammunition drums.

"Let's go Bel, we've got what we came for," he whispered and they melted off through the coconut palms, into the safety of the regrowth and to the place they had secreted their supplies.

Aware of the possibility that the Japanese might discover the theft, they wasted no time in packing. They loaded each of the drums, clipped them to the guns and prepared to set off up the track leading northwards toward the mountains.

"Well mate, we've got four hundred shots to defend ourselves, plus twelve grenades. If we're forced to, we can put up a good fight," said Tal.

Bel's reply sobered him a little, "Let's pray that we do not have to Tal. I've seen enough not to wish killing, even though I've gained hate for their actions. They will have that on their conscience for their lifetime. I do not wish that memory to mar my life. If I am forced to, I shall kill, but not if my enemy is defenceless."

For the rest of the night, they trudged along the well-worn track until about dawn. As they were deciding where to rest, they came upon an adult native, carrying a small crying child. The man was obviously upset, showing signs that he also had been crying.

The trio sat in the bush beside the track and after a lot of questioning in Pidgin English, Bel explained.

"A wounded Australian had made his way to the man's village where soon after he had died. A Japanese patrol had come to the village as the inhabitants were carrying the corpse away for burial. They shot and killed everyone present. The man was alive because he'd taken the child with him to the village gardens."

He implored to be allowed to accompany Bel and Tal over the mountains, but Bel convinced him that both he and the child would be safe by taking refuge in the neighbouring village.

After sharing a meagre meal with him, the two mates left in some haste. The thought of meeting a well-armed patrol held no appeal. Becoming extremely weary after a long walk all day and no sleep the previous night, they moved well into the jungle on the now rising foothills, made a camp of sorts and took stock of their predicament.

Tal's shoulders were by this time aching and raw from the straps of the large pack, which was designed to be hooked in front to webbing attached to the army belt. In it he carried a spare full drum of ammunition, plus two, 100-round cartons, the smaller pack containing the grenades.

The Tommy gun, complete with ammunition, proved to be heavier than he'd imagined. Bel carried his pack filled with the salt and rice and the tins of food given to them earlier plus his gun and his bow and arrows. Tal had endeavoured to talk him into giving these weapons up, but Bel insisted, "No intelligent man would be seen without these weapons in the jungles. They are a badge of manhood in these places particularly. I keep my weapons."

"Fair enough mate. I was unaware of their significance, but the gun will reach a lot further," said Tal.

"With which would you sooner shoot, Tal? The gun would kill you quickly and cleanly. The arrow would kill you just as surely, but you would suffer much before you succumbed. It is a very feared weapon for that reason," said Bel.

"You seem to win all of our arguments," laughed Tal.

"Seriously," said Bel, "it's so good to be free to argue with a white man without fear."

"Thanks for referring to me as a man anyway, I still consider myself to be an addle-brained boy."

Bel ended this exchange with, "Considering your attitude toward me and my people and considering the courage, pain and hardship you have endured, you have proven your manhood. You are no longer a boy, but 'addle-brained'? I'll have to think about that."

In these dangerous circumstances, the heat, mud and rain, this levity between two soul mates cemented further the deep regard each felt for the other. 'Love for your fellow man' became no longer just a phrase from the Bible.

"Tonight, we have a hot meal," announced Bel. He proceeded to forage under a large log, miraculously found wood that would burn and opened a tin of bully beef. The rice, even in a supposedly waterproof bag, was soggy. He broke off a lump of it saying, "It won't take long to cook." He mixed the two together, placed one lump on the smouldering fire and cooked the remainder in the meat tin saying tartly, "We must observe the niceties. Can't have a white man eating ashes."

"Be careful mate. You'll be wearing beef and rice between your eyes," countered Tal, but he bent to Bel's wishes and ate heartily.

Morning came. They took up their burdens and carefully moved out toward the track that would lead them up and over the mountains. Just before reaching the lighter growth a few yards from the track, they were horrified to hear many Japanese voices approaching. They quickly dropped their packs, cocked their guns and waited.

Within minutes a squad of about ten Japanese could be seen through the vegetation, rifles slung carelessly on their shoulders, unaware that two muzzles of death were so close. In a moment they were gone as suddenly as they had arrived. The two mates sank their heads on the ground, suppressing the rush of adrenalin.

"Damn glad they're going back and not our way," whispered Tal.

"Let's hope they're always so accommodating," replied Bel as he wiped the sweat of fear from his eyes.

It was extremely unusual to hear 'drum talk' during the day, but on this day they began. The messages spread out to every village south of the mountains. Bel was able to decipher sufficient to announce, "The people have decided it is now time to break the silence. They tell of the massacre at 'Tol' and of the villagers nearby. They pass on the Japanese message of reprisal for befriending any of their enemies. The most consistent message seems to be 'keep right out of it and let them destroy each other'."

"I was counting on help to obtain food for our journey Tal. It looks as though we may have to steal from their gardens instead," mused Bel.

"Maybe if I stayed hidden, they might give help to you," said Tal.

"We shall see. We still have food for a week if we use it sparingly and we may receive some assistance from mother nature if we are lucky," said Bel.

"There is some bad country up ahead of us and some bad people also. We should now be extremely careful as many people have ventured up here and have never returned. Some of the people live as they have done for countless centuries. They have been known to practice head-hunting and cannibalism and I doubt they will have changed."

This information was to cause a great deal of anxiety to Tal. "I thought that sort of thing died out a hundred years ago," he complained.

"Tal, there are people up here who have never seen a white man. They only know of him through 'drum talk' and what they know they don't like. They are strictly territorial and woe betide anyone passing through without permission and payment. That includes even their neighbours. They are always at war with each other."

"Talk about burning the candle at both ends," said Tal. "We're in the middle of the flaming candle."

"Well put Tal, but not well accepted by you or by me," said Bel. "We shall have to put our trust in God's hands."

"I'll be putting a lot of that in this Tommy gun if it comes to the worst," replied Tal.

HIGHLAND JUNGLE

Knowing the Japanese were behind them and not following, made life a lot easier and the two continued—this time on the track instead of the jungle. Tal considered they were making good time, until on reaching a crest, and in a position where distance could be seen, he beheld a sight that sat him down in disillusionment.

There before them rose mountain upon mountain. The dark green jungle held wisps of mist in their valleys as far as he could see.

Away down to their left, they were able to hear the sounds of a fast-flowing river. According to Bel's knowledge, it flowed down to the sea and emptied out near Tol Plantation. The mouth of this river, one of two rivers, was wide, deep and contained many crocodiles.

"We shall cross many tracks that lead down there," observed Bel. "The people in the hills rely on the river to spear fish, so we shall have to keep a watchful eye in front and at our backs."

They then proceeded to climb the track that wound its way up the ridges. As one ridge petered out, it joined another meandering to the left then to the right, but always upwards.

As the day began to fade, again in the usual pouring rain, they made their rough shelter, had a sparse meal and fell asleep, too exhausted to indulge in conversation. Tal was troubled in sleep, dreaming of the dreadful events of the previous night. The noise of the drums did nothing to mar the sleep of the two unfortunates. Morning arrived, the surrounding jungle swathed in thick and cold fog. They had both lost track of time, not knowing what day it could be, or how long they had been travelling.

"We would probably be only about fifty miles in a straight line from Rabaul, but it seems like we could be at the other end of the earth," was Bel's observation.

"Well, we're still not far enough away," said Tal, amazed that he could find the strength to have endured such a long and arduous trek.

They were now approximately four days from the horrors of Wide Bay and Tol Plantation, and Bel guessed it would be that long again before they reached the top of the mountains, "Providing of course that we do not encounter trouble on our way."

They gathered their belongings and once again struggled up the steep and slippery ridge. The track was narrow, better than the dark and cluttered deep jungle, but they both knew that it was a risk they were forced to take and were at all times every bit alert for any sign of danger.

In about three hours, the fog had disappeared, allowing them to see the blue sky high up through the towering jungle on either side of the narrow track that would have constituted a major thoroughfare for the inhabitants, for perhaps many thousands of years.

There were many birds high in the canopy above them and even though they were in obvious mortal danger, they could not help but marvel at the beauty of this place.

They were surprised to find a large macaw feeding on fallen fruit on the ground in front of them. The bird could not fly, having somehow injured a wing and fluttered along the ground at their approach. Bel put down his gear and dashed back and forth until he caught it. With a deft flick, he broke its neck, smiled his big, bright smile, "Poultry for lunch, Sir."

As it was about that time, they moved off the track, lit a small fire and cooked the bird. The flesh was excellent—the first really solid meat during their journey since they had left the vicinity of Rabaul.

As they left this spot, they received a rude shock. The drum talk began again, but this time they were very close, possibly only a hundred yards away. They stood petrified, knowing there was a village close at hand.

"Let's get going Bel while they're busy making all that noise," said Tal, and they began to hurry as best they could.

Soon they reached a branch in the track. "This would lead in to that village. I think it best we should leave here very quickly," urged Bel. "They were no doubt spreading the news of our presence here."

Before they had taken only a few paces, they were horrified to see about twenty naked and savage-looking kanakas rushing toward them, shaking spears and with arrows strung in their bows. Bel stood his ground, seemingly calm and brave, but Tal quickly cocked the submachine gun and held it at the ready.

Bel called out in an endeavour to make peace and this held them.

They began loud and heated chatter amongst themselves while still making threatening gestures with their weapons.

Bel's attempts were in vain. The chattering altered to become a high-pitched gobbling cacophony and they began advancing with every intention of attack.

Tal fired a burst of just two shots, well over their heads. Some fell to the ground in shock, some turned and ran. The others ceased their awful noise and stood in silence, shocked by the very loud voice of the gun.

Bel walked up to the leader with his hand held up in a sign of peace. He placed his hand on his head, then his heart and did the same to this wild man of the jungle. He then pointed to Tal and himself and then to the mountains.

The man and his comrades then laid their weapons on the ground, had a short rapid conversation among themselves and angrily waved a gesture to be gone.

Looking over their shoulders, the two hurriedly moved off, and the drum beats became more and more active, to be soon echoed by other drums from other places up the mountains.

"Hope I haven't stirred up a hornet's nest with that shooting," gasped Tal.

Bel, equally out of breath, in his haste replied, "If you had not, we would most certainly have been killed Tal. We shall have to wait and see how that will affect others on our way. They have obviously never been exposed to gunfire or they would not have been so afraid. Well done Tal."

From then on, they knew they were being continuously watched, though the watchers were rarely seen. Their presence could be felt and every sound in the jungle caused the hairs to rise on the back of Tal's neck. A pleasant little surprise awaited them that afternoon. After a long five hours of climbing, spurred on by their frightening experience, and the continuing drum talk, they were both in a state of physical and nervous exhaustion. As they rounded a bend in the narrow track, there stood a young spindly and very nervous kanaka boy. He held a basket in his hands and as they approached him, he greeted them with a very shaky smile. He offered the basket and without waiting for a sign of thanks, disappeared into the jungle. The basket contained a hand of bananas, a half dozen five corners and a very generous helping of peanuts.

Bel inspected each piece very thoroughly and declared that none had been tampered with and that they would be safe to eat.

"Well, I'll go to blazes," said Tal. "Looks as though not everyone is out to knock us off."

"Probably some good mother in there somewhere has compassion for us. Thank heavens we were not forced to kill anyone back there or these gifts may have been poisoned," said Bel. "This will keep us going for another couple of days."

They stowed the food in their packs on top of the great hard lump of rice etc. and left the basket where it could be seen. In it Bel placed a sprig of soft, green leaves to show gratitude.

It was becoming late in the day and as had become a wise habit, they set about finding a suitable spot away from the track to make a campsite for the night. They moved about twenty yards into the jungle when Tal remarked, "Bel, this is strange. We've been travelling on top of the ridges and in here the surface slopes only a little bit. Look at the trees. There are no big ones here. What do you make of it?"

Bel agreed it was indeed strange and at the time, could give no explanation.

They searched around and finally found a spot behind a large growth of saplings and vines. There they ate a cold meal of the fruit they'd been given and settled down in their 'gunya' as Tal called it. He then had to explain that their camp would be known as such among the Australian aborigines. After a short discourse on this subject, exhaustion took over and they drifted into silence before sleep.

Tal was shaken back to his senses as Bel whispered, "Tal, do you hear something?" They both listened hard. Tal was listening for voices or rustlings but could hear nothing unusual.

"Listen harder. Can you hear water? Like a stream. A fast-flowing stream. There is no stream away up here. I'm at a loss to understand it," said Bel.

"Yes, I hear it now," said Tal. "Maybe the sound is being carried up the valleys from the river."

"No, I don't think so. Anyhow, we may find out in the morning," said Bel and they gave up and went off to sleep.

Morning came all too soon. Bel was first to stir and Tal saw him wandering close by trying to find the source of last night's mystery. Tal was lying down and he suddenly became aware of the sound again. "Over here Bel. Get down close to the ground and you'll hear it."

Bel did so, and after awhile, crawled toward the centre of the thicket. "Good Lord, Tal. We must get away from here. Don't you come here or we will be gone forever." He eased himself backwards as though he was on the edge of a precipice. "Let's have a look," urged Tal.

"No, most definitely not. This is a very deep hole. Let's get our gear and leave immediately, but tread softly, and watch the ground carefully. I'll tell you all about it when we're out of this."

Bel's urgent voice betrayed fear, causing Tal to follow his example. They walked as though they were walking on broken glass until they reached the safety of the track.

"Now mate, let's hear all about it. It's the first time I've seen you scared," was Tal's great question. "It was an underground river wasn't it?"

"No, my friend, the river was above the 'ground'. We actually slept perhaps one hundred and fifty feet 'above' the ground. We were perched way on top of big old trees that had been swallowed up." Tal listened in silence as Bel went on.

"From down here you can see the growth of vines, mosses and ferns in the treetops. Over many years this growth, in places such as we found, becomes so dense and thick that other trees begin to grow and the original trees die and for a long time support the growth up above. That hole was made because one of the trees on top lost its support from below and crashed through the layer. I was shown a small pocket of this up in the Baining Mountains above Rabaul and to enter beneath it is a chilling, eerie experience. It is inhabited by bats and snakes. The people also believe it is full of the spirits of all the evil ones who died in shame. It is called a 'moss forest' and that is why these tracks, such as this, are on top of the ridges. On real earth."

"Well," sighed Tal, "I'm certainly pleased to have you with me. But it could have a good side as well. What a great place to hide if we had to. It must begin at the top of the ridge because we walked out on it without realising it. If we looked for a pop hole we could hide forever."

"We'll keep it in mind Tal, but you would have to drag me in unconscious. As you know, I am a religious person not given to superscontion, but there must be some of the old ways embedded within me as I don't mind telling you I am deathly afraid of a place like that."

"If it comes to a choice between the moss forest or Japanese or wild kanakas, I'll take the moss forest every time Bel," said Tal as they slung their packs and began another day of heat and rain and climbing.

As they stopped to rest, they took stock of their supplies and found that all that was left was the last tin of bully beef, a tin of condensed milk, a handful of peanuts and about a seven or eight-pound lump of mouldy rice.

"We shall soon have to do something about food Tal," said

Bel. "We have only sufficient for today and tomorrow. I believe our only resort will be to become furtive thieves by resorting to theft from village gardens. We shall have to do this by night, as most of these people are afraid of the dark. Evil spirits roam the dark, so they stay safely in their huts."

They consumed the last of the peanuts and set off once more to conquer the towering mountains that lay between them and the sea at Open Bay.

They stopped again as rain and mist closed in and prepared to find shelter. A voice from the dense undergrowth caused them tremendous surprise. They grabbed for their weapons and dropped to the ground.

"No shoot Masta, me number one friend belong you," came the fast and nervous response.

An oldish black man rose from the undergrowth and showed them that he held no weapons.

Bel shook his hand and Tal offered his. This brought a smile, a black-toothed smile caused by the chewing of betel nut. They allowed the old man to join them as they erected their shelter for the night. They shared the last tin of beef and huddled together in their crude hut of sticks, vines and leaves.

The drums began again as Bel questioned the old man in Pidgin English. It transpired that he had spent many years working on a small plantation along the north coast. His employers had left hurriedly and he had decided to rejoin his relations somewhere down the southern side of the mountains. He had learned pidgin over the years and this was to become a valuable asset to the two travellers. He also understood the 'drum talk' and was to give both good and bad news.

Tal listened hard to the conversation, but could not understand because of the rapid way it was spoken. He had picked up some knowledge of it from Bel, but not at this speed. He was content to sit and listen, knowing that Bel would explain it all.

Bel finally began.

"Firstly, the good news. We have only one more day of climbing and we have reached the top. From then on, it's down all the way. This man has told me that our journey has been closely observed and reported the whole time. The many villagers we have passed, unknowingly most of the time, know that we mean them no harm and prefer not to be involved. They will not prevent us from taking what we need from their gardens, but definitely not whilst they are present—most of them anyway, but unfortunately, not all. Some have stated that they would be safer if they removed us and the danger we present. The very bad news is that a patrol of Japanese is following us quickly and are only a day behind. Another bit of bad news is that the Japanese have already visited Open Bay and Matanakunai, but their two ships and barges have left and headed west toward New Guinea. I think at this time, we should concentrate on Open Bay as we planned and go on from there. He says his name is 'Runi' and that he is well known right down the coast as he worked on boats carrying copra to the plantation at Matanakunai. He says to use his name if we need help."

In the dripping, misty dawn, the duo thanked the old man and quickly left, with the thought of the Japanese on their tails.

"Wonder if they're after us in particular or just checking the place out," said Tal.

"They would most certainly know of us Tal. They would have frightened the information out of someone by this time, so we shall just try and stay ahead of them."

By day's end, they had reached the top. It was almost dark. The track had run out at a long-ago abandoned village. Fossicking about, they had found pawpaws and bananas, a few small sweet potatoes and peanuts. They feasted to the full—the pawpaws flavoured with wild passionfruit. They filled their packs with as much as they could comfortably handle and left. They had to push their way through the dense tangle of undergrowth before they found again the easier going in the jungle.

It was Bel who decided they had reached the top even though everywhere was shrouded in mist.

"Tal, we've made it. There are no more ridges leading up. It's down all the way now."

Happily, it wasn't necessary to build a shelter, as it had not rained all day. They didn't bother with an evening meal, but lounged at the base of a tree and made their plans.

"When we get out of this, I'm going home with you. When they'll have us, we'll join the army together and come back. We'll chase these Japanese right back where they came from," said Bel with great enthusiasm.

"Mate, it would be terrific to have you at home. We could have some good times, particularly out on the farm. We could eat as much as we liked. There are no jungles and no more climbing, and I'm aching to see my family again. We've come a long way from Rabaul. The rest should be easy," predicted Tal, not knowing how wrong his prediction would be.

"How long do you think we've been going Bel," queried Tal. "I reckon it would be over a month now, don't you think?"

"I've just about lost track of time myself Tal, but I'd say we've been going ten or twelve days from Wide Bay. However, we should soon be seeing Kimber Bay. Open Bay is a smaller bay on the north end of it. Going on my memory of the map of our island, I would say we are fairly near."

They set off, downhill at last, chewing raw peanuts for their breakfast. Wasting no time, they slipped and skidded their way down through the dense fog amongst the almost dense darkness of the rainforest. Around mid-morning, the fog lifted and away down over the ridges and valleys lay the ocean.

"We've nearly made it mate," was Tal's joyful yell.

"I hate to dampen your high spirits Tal, but even when we get there we may not be 'out of the woods' as the saying goes. It's still a long walk and could also be dangerous. We shall do well to remain extremely vigilant.

I see smoke down there in a couple of places, so we may encounter villages. We don't know how well we shall be received either."

"You're right as usual mate," said Tal. "By the way, Bel, have you noticed that things are much drier here. I bet there's been no rain here for days."

"That would be right Tal. It's the wet season on the south side at this time of year and it's the dry season on this side. It's the result of varying ocean currents so they say. It probably happens nowhere else on earth that over a distance of twenty miles, two different seasons occur."

"Compared to here, we have a dry season all year round back home," was Tal's contribution. "Anyway, we'd better move. We don't know if the Japanese are still on our tails."

"I think we may have lost them Tal. When the track ran out back there, they would find it very difficult to find which way we went. Let's hope so anyway."

At noon, as they made their way down, they came upon a beautiful series of rock pools, fed by a small stream coming from the steep slope between two ridges. The water was cool, clear and very inviting. Bel studied it carefully, then used one of his arrows to spear about a dozen crayfish, some about ten inches in length. He produced his precious cylinder of matches, lit a small fire and tossed them on.

They made a truly wonderful meal. Tal jumped to his feet, shed his tattered clothing and raced into the water. Bel followed suit and soon, a great game of diving underwater and splashing of each other was in progress. They played as sixteen-year-olds should, in peace, but mindful of their serious predicament. They washed their clothing, gathered up their gear and regretfully left this beautiful place. They had not travelled far, when they came upon another empty village. It had not been abandoned too long ago, as the huts were still in reasonably good condition and the fenced gardens just below contained fruit and vegetables in plenty.

"Now why would they leave such a nice place?" queried Tal.

"Someone in the village has died here, and in order to give the spirit peace, they move away and rebuild on another previous village site. They will return in about six months and re-establish themselves. In the meantime, the garden is ours. We can gather enough for a few days, so let's be at it and be gone," was Bel's advice.

They were soon stocked with pawpaws, bananas, five corners, and some cobs of under-ripe corn.

"Don't forget the peanuts," said Tal and they carried everything to the bottom of the garden.

Here the ground was wet from seepage and in the furrows, they became fairly dirty in the soft black soil.

As they were busily separating the peanuts from the sticky mud, Bel suddenly straightened. "Listen Tal, voices." They listened hard as the sounds of talk and laughter became louder. "They're Japanese. They've caught up with us, and we have nowhere to hide. Quickly Tal, off with your shirt, cover yourself with mud and look as small as possible."

Tal rolled in the black mud and was soon as black as his big mate. Bel smoothed the mud around Tal's shoulders and rubbed what he'd gathered over his mop of curly hair.

"Squat down Tal. You are now my 'Mary'. Keep your gun handy."

Tal cocked both the submachine guns, laid one on his and the other on Bel's feet and they waited. They had only about a half minute and the Japanese appeared. They were moving quickly, no more than fifty yards from the terrified pair. One of the squad spotted them. The leader barked an order and the squad stopped and stared.

"Ten of the mongrels," softly whispered Tal. "If they start up here, we'll get 'em. They've only got rifles."

"Don't do anything yet Tal," was Bel's soft response. "We have only to miss one of them and one of us will die. Just keep still."

The leader yelled something and Bel, standing tall and

clutching his bow and arrows, yelled back a lot of fast gibberish and waved them to move on.

To their complete surprise, the leader barked another order and the squad jogged away down the jungle. As they were leaving, the last two soldiers actually waved happily. They sat back exhausted with relief. "God has been good to us," sighed Bel.

"You're the one who got us out of a mess," said Tal. "If God was looking after us, we'd both be safe at home and there'd be no wars. God's left us mate. It's up to us. Praying is not going to get us anywhere. I'll think about God when this is all over." His voice became angry at this outburst and Bel mirrored his anger.

"You have no right to abandon God. I'll speak no longer on this subject. In fact, I'll speak no more until I've heard you repent your blasphemous conduct. You owe your life to God."

With that he gathered up his gear in silence and moved away. Tal was stunned. "I praised 'him' for saving us and he chewed my head off."

He hurriedly slung his pack, took up his gun and ran after Bel before he could disappear.

Bel moved more slowly now, keeping every sense alert. The Japanese were now ahead of them on a narrow track and he had no wish to catch up with them.

They made camp long before dark this day, giving their enemies plenty of time to get well ahead.

They spoke not a word to each other as they ate their meal and they went off to sleep in separate covers without even making eye contact.

Tal lay there commiserating on this state of affairs that had thrown them apart after such a long hard battle to survive and after such a deep friendship that had lasted for so long. He was so sad that he silently cried himself to sleep.

Bel was later to admit that he had suffered the same emotions.

The morning came, and Tal was the first to speak. "Mate, this can't go on. I'm sorry that I've insulted you and God. You'll have to accept that it was just a reaction to all of this that got to me. I'm ready to accept whatever comes my way without blaming anyone for the cause of it. Will that be a good enough apology?"

Bel, still not smiling replied, "I shall accept your apology for hurting my feelings. It's up to God to accept the remarks you have made."

"That's great Bel. Being offside with you is the last thing I want to see. Cross my heart and hope to die, I promise not to let fly again."

MURDER AVENGED

Feeling better now, they started off again. The track wound along the banks of a small stream and here and there they were able to identify the tracks of the Japanese with their split-toe imprints. They found where they had camped the previous night. The ashes were cold. Obviously, they were in a hurry and had not required a fire.

"I bet they have a date to keep on the coast," said Tal,

"I agree Tal. They are not hunting for us for certain. We shall follow and see what they're up to seeing that they are going our way."

The forest around them had changed. It was no longer dark and gloomy. The canopy was much sparser, allowing sunlight to reach the ground. This, in turn, caused the undergrowth to become a tangled mess of ferns and the dreaded 'wait-a-while' liana vines that climbed right to the treetops. If one blundered into one, the skin would be torn from any exposed parts of the body.

It was late afternoon. The stream was beginning to widen and Bel announced, "That is brackish water and it's tidal. We are very near to the sea and we are also very near to the Japanese."

As if to verify this observation, they heard, just about two hundred yards distant, a chorus of loud hurrahs, shots and laughter.

They decided that the cover was dense enough to creep closer and see what was going on. With all the stealth they could muster, they worked their way along the creek until they were able to observe.

A long barge with a wheelhouse was being driven onto the sand at the mouth of the creek. The two mates counted eleven men as they disembarked amid much talk and laughter.

The tide had begun to rise and they saw one soldier, clad in only a small square of cloth across his front, wade the nearby chest-deep water carrying a machine gun and a belt of ammunition. He set the gun down on its legs in the sand and settled down to keep watch.

"Either our 'friends' are being relieved or being re-enforced," said Bel. "That makes twenty-one of them. I sincerely hope they intend to leave."

"Twenty-one is twenty-one too many. If they stay we'd better take off," said Tal.

The Japanese had all stripped off their uniforms and were clad in similar fashion to the sentry with the machine gun. They began unloading stores from the barge. "Doesn't look good Bel." They then began loading the packs of the first group onto the barge. "Looks good," said Tal.

An extraordinary happening caused the two some consternation. The entire group entered the water, which by now was chest high. They formed a close ring about an apparent officer who began a long harangue. Every now and then his speech was interspersed with loud shouts from the rest of the group.

"Bel, now I'm going to give the orders," said Tal as he took out the smaller haversack he had carried for so long. "We've got them all together and they're helpless. We have twelve grenades here with four-second fuses. We're going to give them to these blokes. Take your six and I'll take mine. We're going to throw them in amongst that lot. Use them like I showed you and get rid of them as fast as you can. Spread them in and around them and I'll bet no one gets out of the water. They quickly, but stealthily moved through the bush until they were hidden by a fold of sand and only thirty feet away from their quarry.

"Now," said Tal. The levers flew off, the springs drove the strikers down onto caps, which ignited the fuse. Four seconds later the grenades began exploding in twos amongst the hapless enemy. The unsuspecting chatter of a few seconds earlier turned to screams of agony.

During the throwing, Tal was able to see the explosions and the devastating effect they caused. Without dwelling on these ghastly happenings, and in this story let it be known that by the time it took to throw his six grenades, he knew his plan was completely successful.

Both he and Bel began an immediate departure, but it was not yet over. They were without their guns and a Japanese soldier had, unbeknown to them, been posted guard on their side of the stream. His head could be seen above the bushes as he ran toward them. As he burst from cover above them, the two felt their time had come. At the same instant, bullets came cracking overhead in a deadly hail.

The hapless soldier fell dead and rolled down the sand to lie twitching and bleeding beside them, killed by his comrade on the other side.

"Got them all but one," said Tal. Think we could wait for him to come out? We could handle him with our Tommy guns."

Bel's answer was to give Tal a great push, rolling him onto his back. "This was a brutal, murderous thing we've just done, and you want to do more!" he raged. "Now I give the orders. We take our gear and leave this place immediately."

"OK, OK, but it felt good to be able to repay them for Rabaul and for Tol Plantation," said Tal.

An odd moan came from the water, but they didn't look. They ran into cover, gathered their gear and headed off down the beach. They stopped exhausted a half-mile away and decided to rest for the night. As they were sitting quietly at the edge of the beach, they heard the sound of a diesel motor. "That machine gunner has a long way to go," said Bel. Bel was frowning and unhappy and though he felt a trifle guilty, Tal felt a thrill of elation. They had won a great battle against overwhelming odds, even if the enemy was helpless. But he told himself, so was the 2/22nd at Rabaul and more so at Tol Plantation. After a meal, he fell back on the warm sand and slept soundly, knowing there was nothing to fear, at least for this night.

Tal was the first to wake next morning. Bel found it extremely difficult to sleep, being very much disturbed about the events of the previous afternoon. He finally awoke. They had a small meal of rice, now fairly mouldy, but it was to prove satisfying after being cooked up in a bully beef tin over a small fire. They were taking their time and about an hour after sunrise, they were startled to hear the distant hum of aircraft.

It wasn't long before they appeared. First two Zero fighters came roaring along above the beach. Next, there appeared four big floatplanes. They circled above the creek where the catastrophe of yesterday had taken place and then proceeded to land on the calm waters.

"They've come to retrieve the bodies," said Bel. "That barge must have had a radio for them to come so soon." "We'd better watch ourselves then. You can bet that if they can carry that many bodies, they are bound to be carrying just as many live ones to take their place."

"Let's get out of here fast," urged Tal.

They gathered their gear and began to run along the sand beside the undergrowth bordering the beach. Very soon they heard the sound of Zeros again. They came flying along the beach, low and very slowly. The two stood stock still and watched them pass no more than fifty yards away. They could clearly see the pilot and he seemed to look straight at them.

Suddenly, the roar became a crescendo as the throttles were opened. They roared up and away to gain height, then banked around and began a dive directly toward the pair. "Quickly, find cover!" roared Bel.

They dived into the undergrowth and cowered behind a small sand dune as the bullets came banging down all around them, covering them with sand and leaves. As the plane roared overhead, they jumped up and rushed further back into the trees. The second plane repeated the attack. This time, they felt safer behind the fairly thick trunk of a tree. Again, the ear-splitting cracking bullets chopped into the trees and sand. A great chip was blown from the tree, a very near miss! They fled again further into the bush as the two planes took turns strafing the areas they had left. They finally exhausted their ammunition and flew away, much to the relief of the two scared but very fortunate victims. As they hurriedly ran away, bursting through the tangle of undergrowth they were able to hear the floatplanes leaving the creek area with their ghastly loads.

They ran as hard as they were able. Both were exceptionally fit and they were able to reach higher ground where the jungle on the hillsides provided more cover and the soft layer of fallen leaves would conceal their tracks. They were aware that their footprints would be easily followed along the beach to their campsite and for some distance inland.

Tal was on the point of collapse with exhaustion. Despite protestations, his big mate grabbed his pack and slung it over his broad shoulders. "We can't afford to stop little friend. We'll just slow down," and they were able to proceed at a steady jog, all the while spurred on by the thought of the consequences should they be caught.

With short intervals of rest, the whole day was spent in this fashion. Dark was not far off when they stopped on the crest of a ridge overlooking the ocean.

"We are now seeing Matanakunai Bay," said Bel. "We've been going all day and have only covered about five miles in a straight line. If we had stayed on the flat land we would have travelled faster, but it would have been much more dangerous.

"I'll guarantee the Japanese are still looking for our tracks," said Tal.

"Do you see what we have ahead of us? At least ten miles of mangrove swamps. To go around would be about three times as far to go. Can you go on Tal? If so, I think we would be safer in there than we would here," said Bel.

"You're the boss mate," said Tal.

Bel decided they would take the short, hard way and they set off. They were fortunate that a mass of logs and debris lay on the mud. They stepped carefully, thus hiding their place of entry into this morass.

It was moonlight and they made slow progress. They had to wade through streams chest high and through mud up to their knees.

"We're lucky that the tide is falling Bel, or we'd be swimming. Hey, what in hell was that?" cried Tal as a loud gurgling roar sounded close by.

"A crocodile. They feed in these channels and lay around on the high spots during the day," was Bel's reply.

"Well, let's find a flaming high spot and spend the night," said Tal. "I don't want them feeding on me. The mossies have already got here."

With that, Bel led the way to a patch of higher ground, which was no more than a patch of high mud, very soft and very smelly.

"Do what I do Tal," and Bel sat down in the mess and coated himself all over, then climbed into the branches of a large mangrove.

Tal handed his pack up to Bel and proceeded to cover himself with that awful mess. "No one will ever speak to me again," brought a chuckle from his mate.

He then began to climb the tree, but at every attempt, he was thwarted by the slippery mud on himself and by that left behind by Bel.

Exasperated, he growled, "Blast the mud, blast the mossies, blast the Japs, blast New Britain, blast this bloody tree." This caused Bel to almost fall from his perch with subdued laughter. "And blast you too, you silly bugger!" raged Tal.

Bel lost his hold, spun around his branch and was forced to plonk down in the mud beside Tal.

"Serves you right for enjoying someone else's troubles," cackled Tal and they thumped each other and laughed till the tears washed the mud from their faces. After recovering from this hilarious episode that drove away all thoughts of misery and danger, Bel hoisted Tal upwards until he was able to help himself, then climbed back up. They sat precariously looking at one another. A flash of white teeth and another chuckle brought "aw shad up".

Tal scrambled out a yard or so where the branch forked. He rolled carefully, put his head and shoulders over the fork and within minutes was fast asleep, so thoroughly exhausted from such a hard day, that the fact he hadn't eaten since early morning, didn't enter his head. Bel hardly slept as he worried that Tal might fall and he marvelled that he did not move a muscle. Upon waking, Tal was also amazed.

"When I tell people about this, no-one is ever going to believe me," he said.

"Well, I'll be there to back you up. Your subconscious must have been wide awake even if you were dead to the world," said Bel. "Here, have some corn. The last of it."

"It's off. It pongs. I'm not going to eat that," said Tal.

"It's not the corn that pongs Tal. It's you, it's me and the mud. Better eat it and we'll be on our way. We'll get a little closer to the beach if we can. That way, we can keep a lookout behind us," said Bel.

"Suits me Bel. Maybe we'll be able to travel faster too."

By midday, they were almost halfway across the swamp. They reached a fairly wide stretch of water, which obviously flowed high and fast during the wet season as there were sandbanks exposed over a good deal of its width.

Two large crocodiles basked in the sun near the mangroves, forcing the pair to make a detour around them and away from cover. They looked back and could see no movement. However, Bel insisted, "You go first and I'll put my feet in your tracks. If anyone sees them, they will think there's only one of us."

"Yeah, I go first and you watch me get eaten by crocs," said Tal.

"Haven't you heard the old saying, 'valour is the better part of discretion'," said Bel.

"I've heard it, but it was the other way around," said Tal as he trudged off.

Soon they were at the sandbanks where they paddled in the water along the edges. As they reached the main stream, they found it surprisingly shallow, barely waist-deep.

"It's fortunate that the tide is low, but it's already beginning to rise. If we are lucky, we shall be out of it by tonight and I won't have to sit and watch you all night," said Bel.

It was well after dark when they emerged from the swamp. They found cover on high ground, dared to light a fire and ate a meal of cooked mouldy rice and half a squashed five corners that Bel found in the bottom of his pack.

They slept well and were woken by an early morning shower. The scene before them was magic. Blue sea, a rocky foreshore and coral.

"This is truly wonderful. Now we shall have plenty of fresh food for a while. Let's go," said Bel with great glee.

They raced down to the shore and Tal was first to rush into the water and luxuriate in the feel of the warm clean saltwater. He dived below and was amazed at the beauty of the coral and clean white sand as opposed to the black sand at Rabaul. Fish of every size and colour swam lazily about the coral.

Bel, on the other hand, was stalking about in knee-deep water with bow and arrows at the ready. It wasn't long before he fired his arrow and happily held up a lovely big parrot fish of about six pounds.

"Here's breakfast Tal. I'm hungry, so let's go and eat it."

This task was soon accomplished and the pair sat back, filled to capacity and feeling very pleased with themselves.

"From this point, we have a great lookout Bel. What do you say we have a spell for the day? We can take turns watching and have a real day off. I'm still dog tired and I'll bet you are too," said Tal.

"I need no coaxing Tal. In fact, I believe they've given up on us for the moment. Yes, I thoroughly agree."

With that, they settled down. During the day, they lazed about, swam, fossicked among the rocks and with Bel's primitive bow, procured more fish for their midday and evening meals.

During the day, Bel warned Tal about the dire consequences of being cut by the coral. "Some of it is quite dangerous. If you're injured, the wounds will form ulcers and without proper treatment, you would be in terrible trouble for a long time"

Tal heeded his advice and the day passed without incident.

HOSKINS PENINSULA

For approximately the next month, the two struggled through swamps, tall cutting kunai grass, and around the foothills of jungle-covered extinct volcanoes. Along white sand and black sand beaches. They procured food from native gardens and fish from extensive coral outcrops.

They saw no Japanese except for shipping and aircraft. The area in which they found themselves was, according to Bel, Hoskins Peninsula and had a fairly large native population.

They attempted to make contact, but as they would approach a garden or village, a loud yell of alarm would send everyone scurrying off into the bush. Frequently, a couple of warriors would stand with weapons at the ready. With rapid-fire language, the two would be warned to move on. Sometimes, they would come across a group in canoes, engaged in fishing. They would not acknowledge Bel's calls but would paddle further out and wait until the disillusioned pair moved away.

"We are not going to receive any help apparently," said Bel. "Either the Japanese have warned them not to, or they have taken sides against us."

They moved on until one afternoon they stood on a rise overlooking a plantation. It was a beautiful scene. Over the great grove of coconuts stood the blue calm ocean.

"See in the distance Tal. The land heads off to the north for as far as you can see. That would be the Talasea Peninsula. We have travelled over two hundred miles. Also, see the people working the plantation. They would have had plenty of contact with white men so we should be able to talk to them. In the morning, I shall approach them."

"I'd rather you didn't Bel," said Tal, "So far we haven't found anyone friendly. We can do without them. Let's camp here and skirt right around them or move along the beach where we can stay in the open."

Tal had no sooner spoken when 'chunk'; a long arrow stuck quivering in the ground about six feet away. The shaft was at a high angle, showing that it had been fired from some distance. They looked hard, but were unable to see any sign of who had fired it."

"That was either a warning or he was hoping to hit one of us Bel. I still don't think it's a good idea to try and talk to them," said Tal.

"We had better stay here for the night then," said Bel. "If we keep moving right now, they could creep up on us without being heard."

They settled down, keeping a good watch all around until after dark. They took turns in trying to sleep, but sleep was impossible. They talked in whispers and decided the beach would be the safer way to travel.

They were on the move before dawn, and upon reaching the beach, jogged steadily past the plantation without incident.

About mid-morning, they came to a place where the foothills of a very high extinct volcano came right to the shoreline. They heard sounds of a village up ahead and Bel declared, "Tal, I'm going up there to talk to them and possibly obtain some food. You wait here for me. While away the time by saying a small prayer for us."

With a grin, he strode off, dismissing protests with one of Tal's phrases, "She'll be right mate."

Bel had been gone for only a minute when a loud yell was heard, plus the barking of a dog. Then the sounds of village life dissolved into complete silence. Even the ever-present songs and chatter of the birds ceased and Tal trembled with a dreadful sense of foreboding.

"Come back Bel," he prayed to himself. He hoped that perhaps Bel was talking to them, which would account for the silence.

Then the silence erupted into an awful, savage, high-pitched gobbling frenzy. They were the war cries that Bel had described in one of their talks and were of the same savage intensity as they had experienced on their previous attempts to make friends. The sounds increased and Tal knew that Bel was in mortal danger.

"Fire your gun Bel. Shoot over their heads. Put some fear into them," he prayed. Amid this awful bloodcurdling clamour, he heard the unmistakable bark of the big .45-calibre bullet from Bel's gun. He knew it was not fired into the air as the shot did not ring out, but was heard as though fired either into a body at close range or into the ground. He decided that the latter would be the case. Knowing Bel's character so well, he could not visualise his big friend turning a gun on his own people.

Listening to the wild clashing of weapons and the rising chanting and screeching, his heart seemed to stop beating. Horror came rushing over him and he felt such shock that his mind refused to function. Bel had died.

He was forced back to reality as he heard excited cries, obviously coming closer and he knew the kanakas were running towards him, down the path from the village. Fear took over and he quickly looked about and considered his options.

He stood stunned and shocked. "What should I do? Should I go up there and shoot them? Should I run? God what can I do?" he cried. "Poor trusting Bel. The best and dearest mate anyone could wish for. Why in hell would they want to hurt him?"

There were no tears as rage, horror and fear took over. He knew beyond doubt that they were coming for him and he had no one to make his decisions but himself.

Where he stood was only twenty yards from cover. The beach to his left offered only the same. Off to his right the beach widened and a spit of sand ran out to about fifty yards. Beyond that the beach ran back close to the woods. He knew now that cover would be of no use to him. The kanakas would use it to their advantage and would no doubt be already waiting.

His reasoning was reinforced as he could hear earnest sounding voices coming from the bush in front of him.

He turned and ran, dropping his pack and sprinted as hard as he could for that spit of sand.

Reaching it, he turned and faced the undergrowth fringing the beach. "If I have to die, this is the place. They'll have to come out to get me and I'll shoot everyone that comes in sight."

He had only fifty rounds of ammunition and he knew it would not be enough. A calmness in the face of death chased away his fear and he waited.

"If I stay here till dark, they might give up. Bel said they didn't like the dark."

With his gun cocked and ready, he waited for the kanakas to show themselves. None of them were brave enough to make a move. As he squatted alone, his body lapped by tiny wavelets, his face became streaked with tears at having to bear the loss of his big, caring black mate. "If he were beside me now, I'd believe there was some hope. I wish I'd gone with him and given him some protection," he recriminated. "Where's God now Bel!" he shouted at the sky through his tears. "He didn't protect you and he's not protecting me!"

He visualised Bel striding confidently into the village, showing his big friendly grin. He would not have fired his gun unless forced to, or maybe it was an involuntary act as he was struck down, unaware of a sneak spear, arrow or ace. Tal hoped that he had died quickly as such a man should not have to suffer after a lifetime of caring for others.

As these thoughts filled his mind, he suddenly became aware of a sound. Aircraft!! Aircraft radial engines. A distinctive sound. "Sounds like at least two of them," he worried as he turned left and right to cover the sky. "I've got kanakas in front of me and Japs behind me. I haven't got a hope," as he visualised 'Zeros' coming in to get him right out in the open.

As he watched the bush and the sky, his eyes were drawn off the beach to his right.

"Good God, it's a flying boat. Twin engines. Must be a Jap patrol. I'm in for it now, but I'll take a heap with me."

As the plane came nearer, he watched it closely. It was only about fifty feet above the glassy water and appeared to take forever to come into clear focus.

Suddenly, he saw it. Red, white and blue roundels on the side. "It's one of ours! It's a Catalina! he roared. "Please see me! Don't go past!" He had a quick brainwave. "This is my only chance. If it doesn't work, I'm dead."

He brought up the machine gun and fired. The bullets raked out across the calm water, churning lines of splashes for a long way out in front of the plane. He watched the big plane fly past and his heart dropped. "They've missed me. What's wrong with them? Now I've got no bullets. I'm done for," he cried.

Things were not as they seemed.

The front gunner had seen what had transpired and radioed to the captain.

"Sir, that was a lone man down there. He was shooting, but not at us. He was trying to get our attention. Must be one of the survivors from Rabaul."

"Right, but first we'll have a look along the beach. Maybe more of them," said the pilot.

They travelled about twenty miles along the coast and saw nothing.

"Now we'll go back and have a look at that chap. Keep your eyes open, all weapons checked and ready for action. It could be a nip trick to lure us down."

Again, Tal heard the thunder of the big motors and soon saw the outline of the big plane approaching. It veered in close and Tal could see all of the guns pointing in his direction. With a great cry of relief, he threw his gun into the water and began to swim.

"It's got to be an Aussie Sir. He's thrown his gun away and is swimming like crazy," said one of the side gunners.

"Right, we'll go round and pick him up. Do not relax. It could still be a trick," warned the pilot.

Tal swam in a panic, visualising hordes of kanakas in hot

pursuit. He began to look about him as he heard those big powerful engines almost bursting his eardrums. He saw the great board wing pass over him and the enormous silver fuselage slide past. Two pairs of strong hands grabbed him and he found himself lying exhausted on the floor between the guns.

"Good Lord, he's only a bit of a kid," said one of the gunners. "What in hell are you doing in this place all by yourself?"

Tal was in a state of shock as the big plane roared, gathered momentum and lifted up and away from his nightmarish brush with death.

Questions were being fired at him, but he was unable to utter a word. As he saw the great peninsula drift away beneath, he knew he was at long last safe.

Overcome, he began to cry. Great heaving sobs wreaked his body as he mourned the loss of his dear friend Bel, without whom he would not have survived. He mourned the loss of all those brave men who had lost their lives back there, and his mind was full of hate and the wish to exact revenge.

At times, he would quieten down, but each time, a kindly word was spoken, he would lapse again into that dreadful sobbing.

"Leave the poor kid alone boys. He'll be OK by the time we get home. Just let him get over it himself," said one of the crew.

Tal was grateful for this man's understanding and within an hour or so was able to accept a cup of warm black coffee and a couple of biscuits. After that, he crouched in a corner, put his head on his arms and silently cried himself to sleep.

He was rudely wakened by a loud boom. He started up, and out of the gunner's 'bubble', was able to see that the plane was in the process of landing.

Two more loud booms as it bounced and it came to rest, turned around and noisily taxied to shore. The engines roared again and it climbed a concrete ramp, turned around and the engines died to silence.

The entire crew accepted his hand in silent thanks. The pilot was last and as he shook Tal's hand he said, "Kiwis to the rescue son. Don't forget your mates from across the Tasman." A jeep drove up and a couple of corporals with large bands on their arms and the letters, M.P. confronted Tal.

"Well, we heard all about you on the radio and we'd like you to come with us. Expect you'd like to clean up and have a good meal eh."

Tal nodded his agreement and asked, "Where am I?"

"Port Moresby my boy. Won't be long and you'll be on your way home, wherever that is."

Tal was led to a barracks where the two military police were very thoughtful and kind. They took him to the showers and to the barber, who gave him a short back and sides. Not altogether too soon as his hair had grown quite long and he needed a shave. He was driven to a Chinese restaurant where he devoured a five-course meal and then to a Chinese shop where he was fitted out with a nice pair of trousers, shoes, socks and shirt.

At the barracks, he was issued with army-issue eating utensils, comb, soap and all of the essentials.

"I should be in uniform," stated Tal. "I joined up in Rabaul and I'm entitled."

"Son, we can't talk about those things. We're only here to see that you're looked after. Tomorrow you'll be able to explain everything. Tonight, we go to the flicks. Forget about everything else in the meantime."

Tal quietly enjoyed the evening, the company, and after nine long weeks, the comfort of a civilised bed.

He went off to sleep and was too tired to be troubled by nightmares or threats of danger. He was truly safe.

GOING HOME

Tal was wakened by his two escorts. Firstly, to the showers, then breakfast of bacon, eggs and toast. They never tasted so good to Tal.

"We go on parade before the C.O. at 0900 hours Tal, so be prepared to answer a heck of a lot of questions," said one of his guardians.

As they had about an hour to go, they wandered about, and Tal was amazed at the enormous amount of supplies and war weapons. Doing all of the work were droves of natives all being shouted at—"boy, do this", "boy, come here".

"Why don't they learn their names? They all have names. One of these days they'll decide they won't take it anymore and there'll be big trouble," observed Tal. This brought no response so he decided to drop the subject.

The C.O. was a kindly man, who listened hard to Tal's story. A corporal sat to one side taking notes. This went on until lunchtime with many, many questions and answers. Tal was relieved when the C.O. finally called a halt and ordered that he would be further questioned by a captain in an adjoining office after lunch.

Tal reported as ordered and very soon became annoyed with the captain. He questioned without regard to feelings, dwelling long and hard on the happenings during the invasion.

He suggested that Tal had behaved very badly as a soldier. "We can't win this war if our soldiers give in so easily. Do you realise that when captured, a soldier is required to give only his name, rank, number and nothing else?" said the captain in a raised and savage tone. "Do you realise that the small amount of information you gave them may have cost the lives of those men at Tol Plantation? To me, your conduct was inexcusable and if it was not for your age and inexperience, you would be shot for cowardice in the face of the enemy."

This brought forth a tirade from Tal, "I'll bet you've never even seen a Japanese. Are you going to win the war with that bloody pen of yours? I'd hate to have had you beside me up there. Why don't you go up there and shake that pen of yours at them? That poor good mate of mine, who died over there was twice the man you are, and you treat them all like dirt. I'd sooner be one of the natives than to be like you and I'll bet you'd have been the first to run."

This sort of exchange went on all that afternoon and all of the following day. At the end of it, Tal was ushered into the C.O., who apologised for the ordeal he'd suffered.

"It has been taken into consideration that in the haste prior to the invasion, you were not able to become aware of the significance and requirements of army law. You were poorly trained and given very little or no instruction regarding these matters. I'm sure you did your very best under the circumstances. I have arranged an honourable discharge owing to your age. You will be paid army wages for the past ten weeks. I have contacted your father at his place of work and assured him of your safety. Tomorrow morning you will be placed aboard a plane enroute to Townsville. From there you will be issued with a rail warrant that will get you home. I wish you all the very best and like you, I hope that this war will soon be over."

The plane was a Dakota. No seats, very, very noisy and as they travelled high over the ocean, it seemed not to be moving at all. Then way below, through open patches in cloud, Tal was to see beautiful sights of the Barrier Reef.

He was dozing as the plane made a perfectly smooth touchdown at Townsville. In fact, he was hardly awake as the engines stopped.

He was collected by another two military police, who drove him by jeep directly to the railway station. They collected his ticket and waited about four hours with Tal until his train was to depart.

The train journey was a terrible experience for Tal. Crammed into a tiny compartment with seven other people, none of whom wished to become friendly or even to engage in conversation.

By the time they reached Brisbane, two days later, all were badly in need of a bath, causing the air to become really foetid.

He shared a taxi to South Brisbane and eventually was aboard the train and heading for home.

After a full twenty-four hours, he began to see the names of places that were familiar to him—Muswellbrook, Singleton, Branxton, Greta, Lockinvar and then the next stop was home. He was about to leave the train when he decided to go on to Newcastle. From there he caught a local train back to where his father worked. He disembarked and waited for over an hour at the tiny station until the train arrived bearing his dad.

As he stood by the gate, he saw his dad's big frame step from the train. His head was down and he was obviously deep in some thoughts of his.

He felt a tug on his arm and a small voice filled with emotion, "G'day Dad."

They stood there for a full minute with his dad holding him close, his big arm crushing Tal to him. Neither was able to say anything. Tal felt a tear fall on his face to join his own. He finally knew it was all over and that he had his life back again.

They walked together to the factory where his dad asked for, and was granted, the day off work. They walked across the road to the jetty on the banks of the river and spent the day going over all that had happened.

As they sat together in this peaceful place, Tal unloaded his complete story. He told of the carnage at Rabaul and Tol Plantation, of the killings at Open Bay and of the loss of his dear friend.

"Dad, the army, at least in the interrogation, made me out to be a traitor because I gave such a small piece of information to the Japs. They would have known anyway as there were planes everywhere. I'm going back there just as soon as I'm old enough to rejoin the army. I don't want anyone to know about all this. I don't think Mum would like to know that I'd killed at my age. Will you promise to keep my secret?"

Patting Tal on the shoulder, his dad made his promise. "A secret it will be until you decide otherwise, but if you ever wish to seek help, just come to me. You'll have to think up some answers to a lot of questions when you get home though. I'll keep right out of it."

"Boy, your mum knows that you'll be home soon, but she doesn't know when. We'll give her a lovely surprise. I didn't wish to worry her and didn't tell her you were at Rabaul. I somehow knew you'd find your way home, but I admit I felt despair at times. I was thinking about you when I stepped off the train. To me, that was a miracle."

"By the way Dad, I've got here exactly the same amount of money that I left with. Here's the amount you gave me plus another pound in interest. I've still got nearly twenty pounds."

As the factory employees emerged after their day's work, Tal and his dad joined them for the walk back to the station and the half-hour trip home.

His mum was busily cooking the evening meal when Tal and his noisy brothers and sister walked in on her.

She too was overcome with joy. After a hearty meal, the questions came thick and fast. "How was Queensland? What did you see? Etc, etc."

He didn't wish to tell lots of lies, so invented stories of what he knew about Queensland.

"Queensland outback is a hot dreary place full of flies and kangaroos," said Tal. "Still, I may go back one day, still a lot to see and do."

As Tal was preparing for bed, his dad came in, placed his hand on Tal's head and for quite a long time, could not find the words he wanted. His voice was trembling and all he could manage was, "I'm proud of you son," and he hurried out of the room.

AFTERWORD

To end this true account, it should be mentioned that he did return to Queensland and then back to New Britain where he served his country well and faced many dangers till war's end.

* * * * * * *

It was April 1942 when Tal was reunited with his family, and it wasn't until August 1943 that he re-enlisted with the army.

He then underwent army intelligence training until January 1945, when he returned to Jacquinot Bay, New Britain and saw action for eight months until the end of the war.

At the end of the war, he went to Rabaul with the army where the disarmament of the Japanese occurred.

At this stage, as soon as time permitted, he went in search of Bel's parents. He could find no trace of them or their village. The village had obviously been abandoned a long time ago, as it was overgrown with grass, vines and trees—just no village.

Neighbouring villagers did not know where they were or were reluctant to tell him any information.

Nobody knew anything.

Tal was honourably discharged from the army in June 1946.

Following his return home in 1946, and after three years of civilian life, which he could not tolerate any longer, Tal joined the airforce. During the next six years within the airforce, Tal served in Malaya during the communist insurgence. But that's another story.

oooOOOooo

www.ingramcontent.com/pod-product-compliance
Lightning Source LLC
Chambersburg PA
CBHW032139040426
42449CB00005B/311